Roger Luiz Da Silva Almeida

Water and nitrogen levels in cocoa cultivation in semi-arid Bahia

Roger Luiz Da Silva Almeida

Water and nitrogen levels in cocoa cultivation in semi-arid Bahia

Irrigated cocoa

ScienciaScripts

Imprint
Any brand names and product names mentioned in this book are subject to trademark, brand or patent protection and are trademarks or registered trademarks of their respective holders. The use of brand names, product names, common names, trade names, product descriptions etc. even without a particular marking in this work is in no way to be construed to mean that such names may be regarded as unrestricted in respect of trademark and brand protection legislation and could thus be used by anyone.

Cover image: www.ingimage.com

This book is a translation from the original published under ISBN 978-620-2-17687-3.

Publisher:
Sciencia Scripts
is a trademark of
Dodo Books Indian Ocean Ltd. and OmniScriptum S.R.L publishing group

120 High Road, East Finchley, London, N2 9ED, United Kingdom
Str. Armeneasca 28/1, office 1, Chisinau MD-2012, Republic of Moldova, Europe
Printed at: see last page
ISBN: 978-620-7-63409-5

Table of contents:

Chapter 1   7

Chapter 2   8

Chapter 3   13

Chapter 4   45

VERY SPECIAL TRIBUTE

To my grandmother Maria, for all the love dedicated to me in this life and certainly in the next.
"In memorian"

DEDICATE

To my wife Marta and my sons Roger and Gustavo for their trust and patience despite the difficulties we've had to face over the years.

OFFER

To my mother Tereza and my brothers Thasso, Yula and Graga for the long years of joy and happiness we had as children and now as adults.

ACKNOWLEDGMENTS

To God Almighty for the teachings for a peaceful and consequently happy life.

To the Postgraduate Program in Agricultural Engineering at the Federal University of Campiña Grande, for offering the opportunity to take the course.

To CNPq, for the doctoral scholarship.

To the Universidade Estadual do Sudoeste da Bahia for allowing me to study for my doctorate.

To Professor Lucia Helena Garófalo Chaves, for her trust, friendship and guidance in the practical and theoretical aspects of this work.

To all the professors of the Graduate Program in Agricultural Engineering at UFCG.

Especially Professor José Dantas Neto, my first teacher.

To the staff of the Irrigation and Salinity Laboratory (LIS).

To my colleagues in the postgraduate course for their friendship and companionship.

Special thanks to my great friend Iedo Teodoro Sampaio for his help and friendship.

To Biofábrica do cacau for the seedlings of the "CCN 51" clone provided for the work.

To Brasmáquinas agricultural implements.

Á Edson Jarade of the Commission for the Elaboration of the Cocoa Farming Plan (CEPLAC).

Á Renilton Santos Santana "Pezao" for his friendship and help with the experimental part.

To Gilney Ferrreira Rios, an agricultural engineer, for providing the area for the experiment.

To Professor and friend Paulo Bonomo from UESB for his friendship and statistical analysis.

The entire Fazenda Velha Irrigation District in Jequié-BA.

To everyone who, directly or indirectly, contributed to the execution of this work.

WATER AND NITROGEN LEVELS IN THE CULTIVATION OF COCOA IN SEMI-ARID BAHIA

SUMMARY

The growth of cacao seedlings and the effect of irrigation and nitrogen fertilization on the growth and productivity characteristics of the cacao tree (*Theobroma cacao L.*) clone CCN-51, was verified in an experiment carried out in a greenhouse and in the field at the Vale do Sol hydro-agricultural property in Jequié-BA. The nursery was made up of 50% shade, inverted micro-sprinklers with a 0.60 m downpipe and an area of approximately 630 m^2 with a 2 m ceiling height. The useful experimental area used in the field was 1792 m^2 planted with clonal cocoa with a spacing of 3.5 x 2.0 m. Four irrigation rates were applied, based on the gross irrigation rate (GL), which were 0.60 GL; 0.80 GL; 1 GL and 1.2 GL; the gross rates were inferred from water evaporation readings in a Class "A" tank installed on the farm. The fertilizer doses applied were 70, 100, 130 and 160 % of the nitrogen (N) recommended for the cocoa crop. The experimental statistical design was a 4 x 4 factorial, in randomized blocks with 4 replications, where the factors water laminae (L) and nitrogen doses (N) were combined, resulting in 16 treatments. In the nursery, plant height, stem diameter and number of leaves were measured; in the field, plant height, stem diameter and production components (number of fruits per plant, length, fruit diameter, fruit weight, number of seeds per fruit, almond weight and yield). The water levels and nitrogen doses used in this study influenced plant height and stem diameter, while the interaction between the water and nitrogen factors did not influence these characteristics. Water laminae, nitrogen doses and the interaction between them had a positive influence on the production of dry almonds and the number of fruits per cocoa plant. In the field, the use of 1922.52 mm of water and 493.10 kg ha^{-1} of nitrogen promoted the highest yield of cocoa beans for sale, 1025.69 kg ha^{-1} . The optimum combination of water table and nitrogen dose was 1926.23 mm and 560.70 kg ha^{-1} of nitrogen, giving a yield of 1649.23 kg ha^{-1} of dried cocoa beans.

KEYWORDS: *Theobroma cacao* L, localized irrigation, fertilization, growth, productivity.

Chapter 1

1. INTRODUCTION

The cacao tree (*Theobroma cacao L.*), of the Sterculiaceae family, a component of a panmictic population native to the Amazon region, is a perennial plant of great economic importance, which generally begins to produce fruit at three years of age. Brazilian plantations face damage caused mainly by biological factors, such as the phytopathological diseases "witches' broom" and "brown rot", caused by the fungi Moniliophtera perniciosa, formerly known as Crinipellis Perniciosa, and Phytophthora palmivora, which occur, respectively, in Bahia and the Amazon, Brazil's largest cocoa producers.

The world's leading cocoa producer is Côte d'Ivoire with a production of 1.5 million tons up to September 2011, followed by Ghana, the second largest producer with an estimated harvest of 1 million tons in the same period. Brazil, once the world's largest cocoa producer with around 40% of this production, is currently in seventh place with an estimated harvest in August 2011 of 237,000 tons (CEPLAC, 2011).

The Brazilian harvest closed in mid-May 2011 at around 200,964.5 tons or 3,349,408 60 kg bags, with the state of Bahia being the largest national producer with 153,393.4 tons or 2,556,556 bags, representing 76.3%; the other states together with 47,571.1 tons or 792,852 bags representing 23.7% of production (CEPLAC, 2011).

One of the new alternatives for recovering cocoa production in Brazil will be the production of irrigated cocoa in the semi-arid northeast. Recent results with irrigated cocoa in the Vale do Sao Francisco show a productivity of 300 @.ha^{-1} , with the plant developing faster than in traditional growing areas (Ilhéus and Itabuna). Another factor that has encouraged farmers and researchers in the region is that productivity can be up to five times higher than that achieved in the south of Bahia, as was also observed by Cruz (2010).

The use of soluble fertilizers in irrigation water, such as urea and potassium sulphate as a source of nitrogen and potassium, respectively, has been one of the practices adopted by cocoa farmers in the semi-arid region of Bahia, who are looking for new technologies to increase productivity.

Siqueira et al. (2011) concluded that fertigation of cocoa crops is a new technology that has proven to be profitable for producers and that soluble formulations have the greatest influence on productivity. According to Borges and Silva (2002), urea is the most widely used soluble source in fertigation, as it has the lowest price and the lowest salt index/unit of nutrient. Malavolta et al. (1989) add that nitrogen stimulates the formation and development of flower and fruit buds and also the vegetative growth of plants, as it is part of the composition of enzymes, coenzymes, vitamins and more than a hundred amino acids, which take part in ion absorption, photosynthesis, respiration, multiplication and cell differentiation.

This new phase of irrigated cocoa cultivation in the semi-arid region of Bahia could be the subject of research under the Plan to Accelerate Development in the Cocoa Region of Bahia - PAC do Cacau, a program created by the state and federal governments that aims to recover cocoa plantations that were devastated by the witches' broom and re-establish the social life of several families in the south of Bahia. Lobao and Setenta (2002) add that the cocoa agro-ecosystem is of fundamental importance under the aegis of sustainable development, as it involves agro-economic, social and environmental aspects of incomparable benefit.

Research into irrigated cocoa in the semi-arid region is beginning to develop, but it is still small in number and its results still need more detailed scientific proof, especially with regard to the crop's field parameters such as the most efficient water table and the most satisfactory nutrient dose that will directly influence the cocoa tree's productivity, as well as other technical aspects that deserve to be diagnosed.

In view of the above, the objective was to analyze the growth of CCN-51 clonal cacao seedlings, evaluate the growth characteristics in the natural environment and the first harvest of CCN-51 clonal cacao subjected to water laminas and nitrogen doses in the semi-arid region of Bahia.

Chapter 2

2. THEORETICAL REFERENCE

2.1. General aspects of the cocoa tree

Cocoa, whose scientific name is *Theobroma cacao L,* is a plant that originated in Central America. *The* Mayan people, located in southern Mexico and Guatemala, were the first to cultivate it. Then came the Aztecs, the first people to use cocoa as a drink. Cocoa was taken to Europe by the Spanish through Christopher Columbus, and they were the first to use cocoa for commercial value. In England, chocolate bars were opened to compete with the traditional English tea. Linnaeus, the father of modern botany, classified cocoa under the name Theobroma, which means "drink of the gods".

The main Brazilian species of cacao tree are the "forasteiro" or purple cacao, *Treobroma leiocarpum, Bern,* and the "criolo", *Treobroma cacao, Linaeus,* the latter being from the *Sterculiaceae* family, native to the South American continent, reaching between 4 and 12 meters in height (Oetterer et al., 2006).

Witches' broom is one of the most serious diseases of the cocoa tree. Originally from the Amazon region, this disease currently occurs in South American countries and Caribbean islands, and is responsible for a loss of around 40% of cocoa production in the Brazilian Amazon and around 30% in Venezuela. In Bahia, it was first detected in May 1989 in the municipality of Urucuca and practically wiped out the state's cocoa plantations.

According to Purdy and Schmidt (1996), the cocoa tree reaches heights of over 15m near tropical regions where the predominant vegetation is of the wooded type, while in other crops this height can vary from 5 to 8m. Cocoa, usually shelled, contains between 30 and 50 seeds and is usually planted with bananas, cassava or other species that provide shade and avoid the effects of wind. A cocoa tree can produce 50 flowers.

According to Namaliu and Daniel (2008), the flowering of cacao trees begins in response to seasonal changes, with hybrid cacao trees starting to flower approximately 30 months after planting, while clonal cacao trees take 15 to 24 months to flower. Full production is achieved when the trees are between four and five years old and can be maintained for 20 years or more with good crop management.

According to Duke (1983), cocoa trees should be shaded for 3 years, removing the flowering shoots until the trees are 5 years old. Cocoa trees should be intercropped with other crops of economic value, such as banana or coconut trees. Irrigation can be used, but always with a drainage system to prevent excess water.

Nakayama et al. (1996), analyzing samples of stem and leaf (clean and petiole), corresponding to the third node, in addition to new leaves 5 cm long, found that the leaf blade is hypo-stomatic, with anomocytic stomata and on both sides of the limb there are four types of trichomes: two tectors and two glandular. The adaxial epidermis contains mucilage-secreting glands, the mesophyll is dorsiventral, the palisade parenchyma is made up of two or three layers and the lacunose parenchyma has collecting cells. Along the mesophyll, there is a predominance of collateral bundles surrounded by a sclerenchymatous sheath that extends to the epidermis, with the venation pattern being of the Camptodroma type with mixed branching Brochidrodomas and Eucamptodromos.

According to Almeida and Valle (2007), cacao produces caulescent flowers that begin to dehiscence in the afternoon and open completely in the early morning following the release of receptive stigma pollen. Unpollinated flowers abscise 24-36 hours after anthesis. The percentage of flowers that pollinate is between 0.5 and 5%. The most important parameters determining yield are related to: (i) light interception from photosynthesis and photoassimilate distribution capacity, (ii) maintenance respiration and (iii) seed morphology and fermentation, events that can be modified by abiotic factors.

2.2. Climatic factors

According to Purdy and Schmidt (1996), the cacao tree produces well at altitudes below 1000 meters and at temperatures below 15°C it slows down its growth. At temperatures of 5°C, growth stops and survival can be affected, and a short exposure to 0°C can kill the plant. On the other hand, high radiation and temperature increase plant metabolism and, consequently, increase the requirement for water, nutrients and cultivation. The effects of light and nutrients on cocoa plantations are closely related. Thus, the establishment of shade for cocoa trees and the response to fertilizers cannot be separated (Muller and Biehl, 1993).

Burridge et al. (1964), carrying out chemical analyses of leaves over a two-year period on cocoa trees in a field fertilized with shading and irrigation treatments, found that shading increased the levels of nitrogen, phosphorus and potassium, and decreased the levels of calcium in the leaves.

The plants are shade-tolerant and require evenly distributed high temperatures, with a maximum daily temperature of 33.5°C and a minimum of 13°C, with a daytime temperature range between 33.5 and 18°C. They are intolerant of wind, which is why they are often planted on slopes to protect against wind (Reed, 1976).

Regions with high rainfall or well-distributed rainfall throughout the year provide favorable conditions for the occurrence of witches' broom (Luz et al., 1997), adding that Purdy and Schmidt (1996) also consider that the microclimate provided by the plantation favors the infection and sporulation of cocoa by the Crinipellis perniciosa fungus.

According to Gagné (2008), the cocoa tree is subject to many diseases such as fungi, pod rot and strange growths and low temperatures kill the seeds. The cocoa tree requires year-round moisture for its growth and regular irrigation is necessary. Cocoa is a very needy and demanding plant, and if these needs are not met the plant will die.

According to Hardy (1961), Alvim (1972), Augusto (1997) and Castro and Kruge (1998), the cocoa tree's climatic limitations for production are: average monthly rainfall of at least 100 mm in the driest month, average monthly temperature of at least 22°C and an absolute minimum temperature of no less than 6°C; according to Duke (1978), the cocoa tree can withstand an annual rainfall of 480 to 4290 mm and an annual temperature ranging from 18.0 to 28.5°C,

In order to produce cocoa in regions with a dry climate, it is necessary to adopt irrigation systems. These regions have rainfall ranging from 600 to 800 mm per year, distributed over the months of December to March, while the cocoa tree needs 100 to 150 mm per month to obtain good yields (Gramacho et al., 1992).

In addition to the water supply, macro and micronutrient fertilization should be carried out via irrigation water so that the plants are well nourished and thus better able to express their genetic potential, as well as making them more tolerant to pests and diseases (Mi, 2006).

2.3. Cocoa seedlings

Marrocos and Sodré (2004) state that the large-scale production of clonal cocoa seedlings began in 1999 at the Instituto Biofábrica de Cacau (IBC) in Bahia.

Clonal or seminal cocoa seedlings, irrigated in regions with temperatures above 35°C, have shown resistance to witches' broom, since thickening of the apical bud and twisted leaves with swollen pulvinus, the main indicators of the disease in seedlings, have not been observed in nurseries (Alves and Del Ponte, 2010). According to these authors, although the ideal conditions for the development of basidiocarps are annual rainfall of between 1,500 and 2,000 mm, temperatures of between 24 and 26°C and relative humidity of between 80 and 90%, seedlings in the adult stage, subjected to irrigation, regular fertilization and good plant health care, can produce satisfactorily, as has been shown in some regions of Brazil.

2.4. Cocoa tree growth

The cocoa tree requires year-round moisture for its growth and regular irrigation is necessary (Gagné, 2008). According to Purdy and Schmidt (1996), the height of the cocoa tree can vary from 5 to 8m, with rainfall ranging from 1250 to 2800 mm per year. Arce (2004) and Gramacho et al. (1992) cite similar rainfall values for maximum plant growth.

According to Almeida and Valle (2007), the growth and development of the cacao tree is dependent on temperature, which mainly affects vegetative growth, flowering and fruit development. Waterlogging of the soil reduces leaf area, stomatal conductance and the rate of photosynthesis, as well as inducing the formation of lenticels and adventitious roots. For most genotypes, drought resistance is associated with osmotic adjustment,

For Peixoto and Peixoto (2004), quantitative growth analysis has been used by plant researchers in an attempt to explain differences in growth, whether genetic or the result of environmental modifications. Benincasa (2003) points out that growth analysis makes it possible to assess the final growth of the plant as a whole and the contribution of the different organs to total growth, allowing us to learn about functional and structural differences between cultivars of the same species in order to select them and better meet the desired objectives.

Wilhelm and McMaster (1995) define growth as the irreversible increase in the physical size of an individual or organ over a given period of time.

2.5. Cocoa tree irrigation

In order for the plant to develop properly and obtain satisfactory yields, it is essential that water and nutrients are replenished in the ideal quantity and at the right time (Nanetti et al. 2000). The response of crops to different combinations of irrigation and fertilizers has been the subject of numerous studies (Frizone et al., 1996).

However, according to Dias and Resende (2001), research with cocoa trees in experiments requires an excessive amount of time and area, representing high costs in terms of implementation, management, harvesting and labor. For these reasons, the simplicity, flexibility and robustness of the experimental design to be used and the multiplicity of research objectives are important aspects that must be observed in experimentation.

Localized irrigation in cocoa cultivation is starting to become commonplace in modern cocoa plantations. Areas of cocoa irrigated by drip irrigation or micro-sprinklers can already be found in various parts of South America. The trend of cocoa cultivation away from forests and towards open fields is being accompanied by the introduction of localized irrigation technology (Netafim, 2008).

An irrigated cocoa plantation, with a spacing of three meters between plant rows (1,100 plants per hectare), produces 1500 kg ha^{-1} , this same plantation without irrigation achieves an average productivity of 600 kg ha^{-1} , when the plantation is irrigated and fertilized, productivity can double. From the 1500 kg ha^{-1} achieved on average with irrigation, the producer who uses fertigation can achieve an average of 3000 kg ha^{-1} . As well as increasing yields, irrigation increases seed weight by up to 70%; an unirrigated crop produces seeds weighing 0.80 grams on average. Irrigated crops produce seeds with an average weight of 1.40 grams (Siqueira, 2008).

In the north of Espírito Santo, irrigated cocoa trees had an increase in production of 54% (Siqueira et al., 2011), while in the Reconcavo Baiano, production increased by up to 100%. The feasibility of implementing irrigated cocoa cultivation in the semi-arid region should be realized when research indicates varieties adapted to the region's climatic conditions. According to CODEVASF (2009), in the Vale do Sao Francisco, using new technologies such as fertigation and drip irrigation, producers are expecting to achieve production of up to 300 arrobas.ha^{-1} of cocoa (4.5 tons), while in the traditional region they don't reach 50 arrobas.ha^{-1} (0.75 tons).

2.6. Nitrogen fertilization for cocoa trees

According to Chepote et al. (2005), cocoa tree fertilization is based on the doses of nitrogen determined in field trials and the critical levels of phosphorus and potassium available that provide the greatest development and production of the cocoa tree; however, Morais et al. (1979) state that nitrogen, in addition to being a nutrient that greatly limits the growth of cocoa seedlings, can be lost through volatilization and, in an environment with constant irrigation, suffers high losses through leaching.

Nitrogen is one of the factors that causes the greatest morphophysiological changes in plants, with the possibility of altering the number, weight and quality of fruit. It is essential for the synthesis of amino acids, chlorophyll, alkaloids, nucleic acids, hormones, enzymes and vitamins (Marschner, 1995).

Burridge et al. (1964) studying the effect of fertilizers and irrigation on cocoa tree leaves found an increase in the levels of nitrogen, phosphorus, calcium, and magnesium, and a decrease in the level of potassium and the effect of irrigation was small, but decreased the levels of nitrogen, potassium, calcium, and magnesium.

Vernon et al. (1972), applying fertilizers in the third and fourth year of the cocoa tree, found no significant response to fertilizer application, even though the plantation had a yield of 3000 kg ha[-1].

Zobel et al. (2007) concluded that the diameter of cocoa tree roots changes in size due to the concentration of nutrients, increasing or decreasing according to the concentrations of nitrate, phosphorus and aluminum. The change in diameter depends on the type of nutrient, the seed variety and the interaction between them.

The primary macronutrients nitrogen and potassium have been the most widely used in ammoniacal fertilizers. They are highly soluble, have a high salinity index, a high acidity index and often lack secondary macronutrients (Borges and Silva, 2002).

2.7. Crop management

In order to achieve higher yields for the cocoa crop, different spacings have been used in the field and the best form of propagation has been investigated. Mooleedhar and Lauckner (1990) investigated the effect of three spacings for the cocoa crop: a traditional planting of 3.6 x 3.6m (748 plants ha^{-1}), an intermediate planting of 3.6 x 1.8m (1495 plants ha^{-1}) and a personal planting of 1.8 x 1.8m (2990 plants ha^{-1}). The yields obtained by the last spacing were significantly better than those of the intermediate and traditional spacings.

According to Duke (1983), spacings of 2.4 m x 2.4 m or 3.6 m x 3.6 m should be used, but if the soils are poor and at altitudes above 300 m it is advisable to use smaller spacings.

According to Duke (1983) cocoa propagation can be done by grafting, but sowing is a less costly procedure. Recent work by researchers into cocoa propagation systems emphasizing genotype multiplication and gathering a collection of seeds for distribution has had more effective results than the grafting procedures used worldwide to propagate the crop, since seeds are a limiting factor in production due to the fact that most cocoa genotypes are highly heterogeneous (USDA, 2003).

The cocoa tree should be intercropped with other crops of economic value, such as banana or coconut trees, which should remain shaded for 3 years, removing the flowering shoots until the trees are 5 years old.

2.8. Cocoa tree productivity

According to the International Cocoa Organization (ICCO), net world production for the 2010/11 harvest was 4.2075 million tons (t), derived from gross production of 4.250 million tons, far exceeding the 4 million ton mark for the first time in history, with an apparent additional increase of 619 thousand tons, or 17.0%, compared to gross production of 3.631 million tons in the 2009/10 agricultural year, CEPLAC (2011).

The high yields shown in the research carried out in recent years by the Executive Commission for the Cocoa Farming Plan (CEPLAC) with irrigated cocoa, positively signal a probable paradigm shift, through the acceptance of the hypothesis that irrigated cocoa farming is a profitable activity (Begiato et al., 2009).

The efforts of research and technical assistance in Espirito Santo led to a 59% increase in the productivity of dried cocoa beans between 1955 and 1985, for an expansion rate of less than 10% (Smith, 1990). However, in order for the plant to develop properly and obtain satisfactory productivity, it is essential to replenish water and nutrients in the ideal quantity and at the right time (Nannetti et al., 2000), 2000). On the other hand, the type of seed, among other factors, is a limiting factor in production due to the fact that most cocoa genotypes are highly heterogeneous (USDA, 2003).

Zuidema et al. (2005) state that the cocoa production models that have been established to date are based on regression with limited applicability to locations other than the research site, simulation models can be valuable for identifying gaps in knowledge about cocoa production. For perennial crops, few studies have presented production models, while for temporary crops there are more studies on the subject (Cannell, 1985). Teal and Vigneri (2004) analyzed the evolution of cocoa production growth in Ghana using a production function that considers inputs and land as independent variables.

2.9. Production plant

The existence of various factors related to climate, soil, plants and others interacting with each other determines the productivity of agricultural crops, i.e. there is a mathematical relationship between these factors and productivity that can be expressed by:

$$Y=f(x_1,x_2,...,x_n) \tag{1}$$

Where:

Y = quantity produced.

$x_1, x_2, ..., x_n$ = inputs used for production.

According to Frizone and Andrade Junior (2005), economists define a production function as the physical relationship between the quantities used of a certain set of inputs and the maximum physical quantities that can be obtained from the product, for a given known technology. Production functions are widespread and are often used to determine the optimum or economic levels of factors of production,

The exploitation of any agricultural activity aimed at obtaining a product requires the use of a certain amount of resources, which are combined in quantity and quality according to the knowledge of available technologies, by those who decide to carry out the exploitation. The relationship between the process of converting various factors of production (resources) into a given product is a production function, whose input-output relationship can be continuous or discontinuous (Aguiar, 2005).

The mathematical models that describe the production function most commonly used in the economic analysis of agricultural research are: quadratic, square root, Mitscherhich and the 3/2 power, Hexem and Heady (1978). However, when working with a greater number of levels of the factors considered in the production analysis, the quadratic type relationship is the most commonly used, i.e. when we have a relationship with two factors, a second degree polynomial is obtained and a three-dimensional response surface can be the graphical representation of the function described by the polynomial.

Simulation models are also very important tools in the study of soil-crop-climate interaction, since they make it possible to consider a large number of environmental factors that affect the crop, as well as to analyze soil effects, which would be impossible in conventional experiments, due to the high costs and long time required to obtain research results. The use of models therefore saves time, financial and human resources.

2.10. Water use efficiency (USA)

According to Oliveira (1993), the relationships obtained between the amount of water applied and the water use efficiency (WUE) allow us to know how the plant is using water in its process of transforming it into a marketable product.

Klar (1988) states that soil fertility, in particular, promotes greater efficiency of water use by crops, with nitrogen being one of the nutrients that promotes significant variation in the efficiency of water use by crops. According to Lopes (1989), when the yield of a crop increases with fertilization, the efficiency of water use by the crop also increases.

The efficiency of water use is determined by the relationship between the yield obtained by the production function as a function of the total amount of water and the levels of nitrogen applied, and the total amount of water applied (Doorembos and Kassan, 1994).

2.11. Nitrogen use efficiency (NUE)

Due to the various types of losses, the efficiency of nitrogen use by plants is around 40% to 50% (Sartori, 2010), especially urea and other inorganic fertilizers which are lost through ammonia volatilization and can reach up to 55% of the nitrogen applied (Alkanani et al., 1991). However, alternatives have been sought to reduce these losses through more efficient management of the application equipment and the amount applied in the field.

According to Muraoka and Trevelin (2007), nitrogen use efficiency, measured by grain yield per unit of nutrient applied, should be achieved in high-yielding crops with good management practices. According to these authors, nitrogen use efficiency can be expressed through the internal nitrogen use efficiency, the partial productivity factor, physiological efficiency, agronomic efficiency and apparent recovery efficiency.

Chapter 3

3. MATERIAL AND METHODS

3.1. Characterization of the experimental area

The research was carried out on the Vale do Sol hydro-agricultural property from July 1, 2009 to February 28, 2012, in Jequié, Bahia (13° 51' 28" S , 40° 5' 2" W and altitude 199 meters), approximately 300 m from the stone dam of the Rio de Contas (Figure 1). The Koppen climate classification indicates that the research area is under the Aw climate: a warm climate with the coldest month having an average temperature of over 18 °C. The average annual temperature is 23.6 °C and in the summer at the site of the experiment, the temperature reaches 45 °C.

The minimum relative humidity is 58.3% and the maximum 72.9%. The average annual potential evapotranspiration is 1,500 mm, with a maximum in December and a minimum in June, with indices of 169 mm and 68 mm, respectively. The winds have a low annual average speed of 3.6 Km/h, a maximum of 5.8 Km/h and a minimum of 2.2 Km/h. The city of Jequié is located in the southwest region of Bahia, 365 km from Salvador. The Jequié microregion is one of the microregions of the Brazilian state of Bahia belonging to the Centro-Sul Baiano mesoregion. It has a total area of 17,396.126 km^2 .

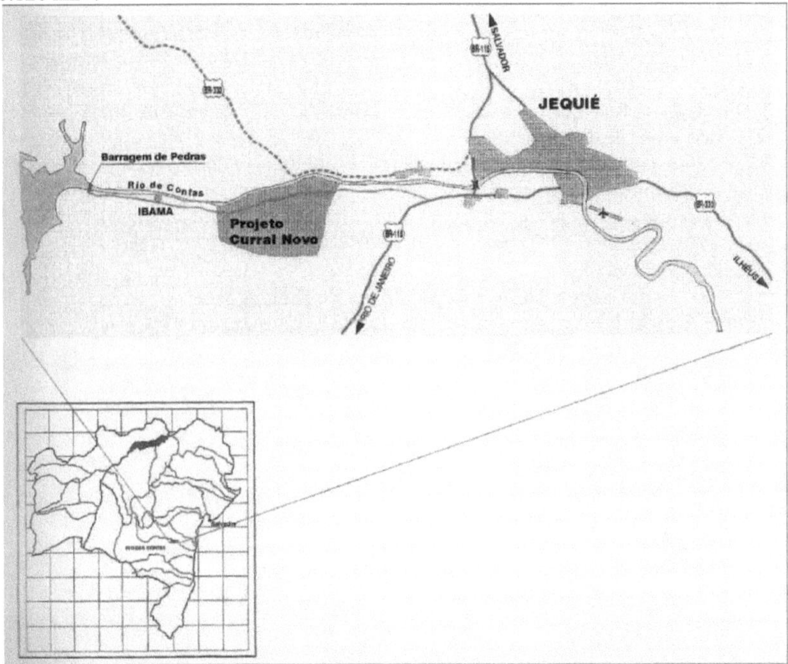

Figure 1 - Location of the experimental area. Jequié-BA, 2012.

3.1.1. Protected environment

The farm's nursery is made up of 50% shade, dampers, 1" PVC pipes (25 mm) and ^" inch hoses (24 mm) which supply the inverted micro-sprinklers with a 0.60 m downpipe. Its area is approximately 630 m^2 with a ceiling height of 2 m (Figure 2).

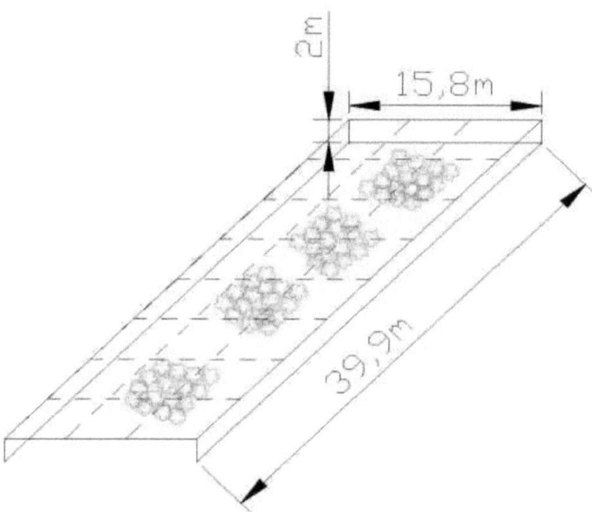

Figure 2. Dimensions of the nursery used in the experimental area. Jequié-BA, 2012.

3.1.1.1. CCN-51 clonal cacao seedlings

The 180-day-old cuttings obtained by cuttings from the Biofábrica do Cacau, located in the municipality of Uruguca-BA, were placed in black polyethylene bags and kept under 50% shade (Figure 3). The soil material had the following physical-chemical characteristics determined by the method adopted by Embrapa (1997): sand = 700.3 g kg^{-1} ; silt = 232.6 g kg^{-1} ; clay = 67.1 g kg^{-1} ; pH (H2O) = 7.83; Ca = 8.92 cmolc kg^{-1} ; Mg = 2.81 cmolc kg^{-1} ; Na = 0.31 cmolc kg^{-1} ; K = 0.40 cmolc kg^{-1} ; H + Al = 0.00 cmolc kg^{-1} ; MO = 25.5 g kg^{-1} ; P = 56.2 mg kg$^{-1.}$

Figure 3: Distribution of seedlings in the nursery. Jequié-BA, 2012.

3.1.1.2. Growing the seedlings

The crop treatments carried out were sprinkler irrigation, manual weed control and phytosanitary control.

During the experimental period, the seedlings were sprayed with a cocktail consisting of chemo-oil 0.5 mL L^{-1} of water, talfontop 1.5 g L^{-1} of water and Aminon organomineral foliar fertilizer 2 mL L^{-1} of water, split into three applications.

3.1.1.3. Irrigation of seedlings

The irrigation system was installed with inverted micro-sprinklers, with a 0.6 m long downpipe, with a flow rate of 48.6 L h^{-1} for a pressure of 1.3 kgf cm^{-2} . The irrigation system ran from 7:00 to 8:00 am and from 5:00 to 6:00 pm daily.

The Cristiansen Uniformity Coefficient (CUC) (1942) was calculated to estimate the water distribution uniformity of the micro-sprinklers used in the nursery (Equation 2). The averages of the volumes (mL) of water collected in the containers spaced 1.5 m apart were used. Three tests were carried out, with a set time of L h-1 for a single micro-sprinkler working with a working pressure of 1.3 kgf cm-2 and an average flow rate of 48.6 L h^{-1} , with a CUC value of 81%.

$$CUC=100.\left[1-\frac{\sum_i^n|q_i-\bar{q}|}{n.\bar{q}}\right] \qquad (2)$$

Where:
q_i = flow rate at each emitter, Lh^{-1} ;
q = average emitter flow, Lh^{-1} ;
n= number of emitters.

3.1.1.4. Substrate

Substrate was used, based on soil with the physical and chemical characteristics described in 3.1.1.1 and goat manure in a 4:1 ratio, fertilized with 400g of simple superphosphate (P2O5 16%; S 11% and Ca 16%) and 16g of FTE BR-12 (9% Zn; 1.8% B; 0.8% Cu; 2% Mn; 3.5% Fe; 0.1% Mo).

3.1.1.5. Variables analyzed

During the period from 19/06/2009 to 21/11/2009, the CCN-51 cocoa seedlings remained in the nursery where three growth parameter evaluations were carried out according to the schedule of dates (Figure 4): plant height from ground level to the basal part of the last leaf measured with a tape measure, stem diameter at the soil surface measured with a 150 mm steel caliper to an accuracy of 0.05 mm and the number of leaves (direct count) (Figures 5 A and B).

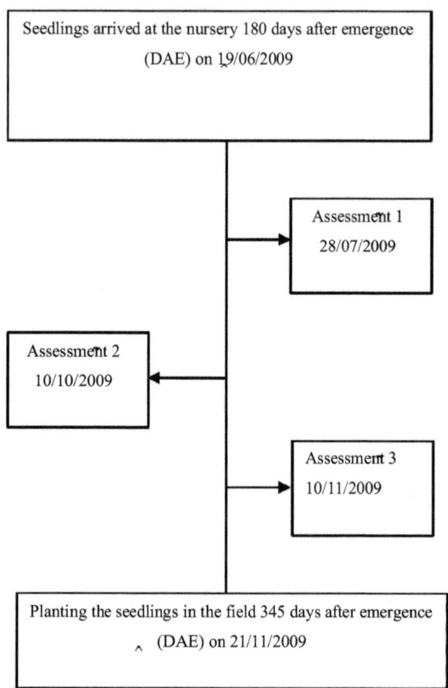

Figure 4 Organizational chart of the evaluations carried out in the protected environment. Jequié-BA, 2012

Figure 5 Measurement of stem diameter (mm) (A) and measurement of height (cm) of CCN-51 cocoa seedlings (B). Jequié-BA, 2012.

3.1.1.6. Statistical analysis

Statistical analysis was carried out using descriptive statistics and the Tukey test using the ASSISTAT 2011 program and Excel 2007 to obtain the averages of the characteristics evaluated.

3.1.2. Natural Environment

From 21/11/2009 to 29/01/2012, clonal cacao CCN-51 was planted in 0.40 x 0.40 x 0.40 m holes, 3.5 x 2.0 m apart, in the experimental field area of approximately 0.18 hectares, with slightly undulating terrain (Figure 6).

Figure 6 - Area demarcated with a square. Jequié-BA, 2012.

3.1.2.1. Climatic characteristics

The climate data, evaporation and precipitation, were obtained from daily readings taken at 7:00 am in the class "A" tank (Figure 7A), with a tranquilizer pogo and glass ruler graduated in mm and in the Vally de Paris rain gauge, installed in the mini weather station located in the experimental area, while the temperature and relative humidity data were provided by the weather station of the Fazenda Velha irrigation district located approximately 100 m from the experimental area (Figure 7B).

Figure 7 Mini weather station in the experimental area (A) and weather station in the Fazenda Velha Irrigation District (B). Jequié-BA, 2012.

The average reference evapotranspiration in 2010 reached a maximum of approximately 160 mm in November and a minimum of 90 mm in August, always exceeding the monthly rainfall, which reached a maximum of around 80 mm in December, requiring the use of supplementary irrigation (Figure 8). In 2011, the average reference evapotranspiration was below 160 mm, with average rainfall above 0 mm for practically the whole of 2011, peaking in November with excess water in the soil of approximately 20 mm (Figure 9).

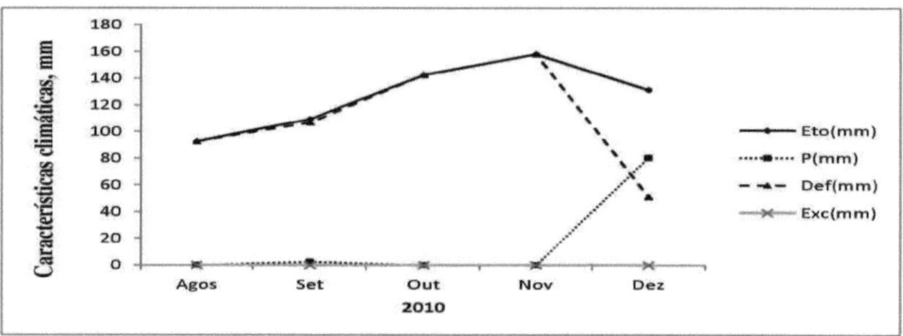

Figure 8 Water balance during the experiment period in 2010, where Eto = reference evapotranspiration; P = rainfall; Def = water deficit and Exc = excess water. Jequié-BA, 2012.

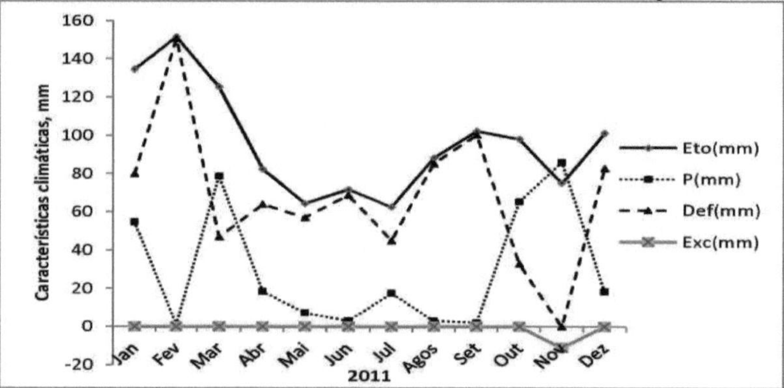

Figure 9 Water balance during the experiment period in 2011, where Eto = reference evapotranspiration; P = rainfall; Def = water deficit and Exc = excess water. Jequié-BA, 2012.

Cocoa trees require a warm climate to avoid infestation by witches' broom. Temperatures varied significantly throughout the experimental period, reaching highs of 37°C and lows of 14°C in 2011, with the average temperature showing a uniform trend of around 27.5°C (Figure 10). The average relative humidity was around 61%, making it difficult for the fungus Moniliophtera perniciosa, which causes witches' broom, to appear on the cocoa crop.

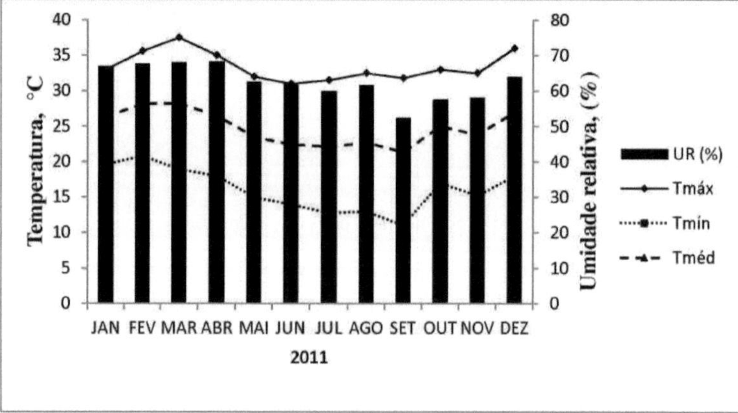

Figure 10 Climatic characteristics for 2011 in the experimental area, where RH = relative humidity; Tmax = maximum temperature; Tmin = minimum temperature and Tmed = average temperature.

Jequié-BA, 2012.

3.1.2.2. Characteristics of the soil and water used for irrigation

The physical and chemical analysis of the soil samples was carried out at the Irrigation and Salinity Laboratory of the Agricultural Engineering Department at the Federal University of Campina Grande. According to the results in Table 1, the experimental area was fertilized with potassium and phosphorus as recommended by CEPEC/CEPLAC (2009).

According to the values shown in Table 1, there was no need to add dolomitic lime to increase base saturation, as it was 84.69% higher than the recommended 50% for growing cocoa. The analysis also revealed that there was no toxic aluminum due to the pH above 5.5 present in the soil particles, indicating a soil without acidity problems. Fertilization was carried out on the basis of the soil analysis (Table 1) and CEPLAC's recommendations for growing cocoa.

Table 1 - Physical and chemical characteristics of the soil in the experimental area. Jequié-BA, 2012.

Features	Depth soil		
	(0-0,20 m)	(0,20-0,40 m)	(0,40-0,60 m)
Sand (g kg)$^{-1}$	683,8	645,3	634,0
Silt (g kg)$^{-1}$	228,8	227,7	218,4
Clay (g kg)$^{-1}$	87,4	127,0	147,6
Texture	Sandy loam	Sandy loam	Sandy loam

Soil density	1,57	1,54	1,64
Particle density	2,76	2,73	2,76
Porosity (%)	43,12	43,59	40,58
Natural humidity (%)	0,70	0,20	0,55
Field capacity (%)	10,63	9,66	11,13
Wilting point (%)	3,63	3,47	3,88
Available water (%)	6,97	6,19	7,25
Calcium (cmolc kg)$^{-1}$	2,53	2,32	2,07

Magnesium (cmolc kg)$^{-1}$	1,81	1,76	1,98
Sodium (cmolc kg)$^{-1}$	0,03	0,04	0,07
Potassium (cmolc kg^{-1})	0,28	0,12	0,1
S (cmolc kg)$^{-1}$	4,65	4,24	4,22
Hydrogen (cmol$_c$ kg-1)	0,84	1,02	0,9
Aluminum (cmolc kg)$^{-1}$	0,00	0,00	0,00
T (cmolc kg)$^{-1}$	5,49	5,26	5,12
Qualitative calcium carbonate	Absence	Absence	Absence

Organic Carbon (g kg)$^{-1}$	0,58	0,58	0,25
Organic matter (g kg-1)	1,00	1,00	0,43
Nitrogen (g kg-1)	0,05	0,05	0,02
Assimilable phosphorus (mg kg)$^{-1}$	11,2	3,4	2,2
pH in H2O (1:2.5)	6,48	6,51	6,61
Electrical Conductivity (dS m)$^{-1}$	0,13	0,12	0,12

  The chemical analysis of the water used for irrigation was carried out by SENIR- DNOCS (4ª diretoria regional de estudos e projetos) located in Salvador-BA, with the characteristics shown in Table 2.

Table 2 - Chemical characteristics of the water used by the irrigation system in the area experimental. Jequié-BA, 2012.

Features	**Value**
pH	6,9

Electrical Conductivity (dS m)$^{-1}$	0,24
Calcium (mmolc L)$^{-1}$	1,0
Magnesium (mmolc L)$^{-1}$	1,24
Sodium (mmolc L)$^{-1}$	0,87
Potassium (mmolc L)$^{-1}$	0,46
Chlorides (mmolc L)$^{-1}$	1,73
Bicarbonates (mmolc L)$^{-1}$	1,70
Sulphates (mmolc L)$^{-1}$	Absent
Sodium adsorption ratio (SAN) (mmol L) /$^{-112}$	1,3
Water Class	C2S1

Table 2 shows that the water used for irrigation was classified as C2-S1, i.e. water with medium salinity and low sodium content that can be used for irrigation on almost all types of soil; the electrical conductivity of the irrigation water of 0.24 dS m^{-1} and 2 dS m^{-1} estimated for cocoa were used to calculate the leaching rate.

3.1.2.3. Spraying and cleaning of the experimental area.

Throughout the experimental period, spraying and cleaning of the area was carried out according to the needs presented in the area, mainly combating existing weeds, either by manual cleaning or with the use of a brushcutter, as shown in Table 3.

Table 3. Spraying and cleaning of the area throughout the experiment. Jequié- BA, 2012.

DATE	DEFENSIVE	QUANTITY/TYPE	CLASS
07/08/2010	RECONIL	4 g.ft.$^{-1}$	Bactericidal fungicide
25/01/2011	DECIS 25Ec	200 ml.ha^{-1}	Contact insecticide
30/07/2011	DECIS 25Ec	150 ml.ha^{-1}	Contact insecticide
CLEANING THE AREA			
10-1112/11/2009		Manual	
26-27/03/2010		Manual	
15/06/2010		Brushcutter	
30/09/2010	Roundup Multi-action	2 kg.ha^{-1}	Non-selective herbicide
08/01/2010	Podium EW	0.75 l.ha^{-1}	Selective herbicide
20/04/2011		Brushcutter	
29/08/2011		Brushcutter	

| 04/11/2011 | | Manual | |
| 09/01/2012 | | Manual | |

3.1.2.4. Pruning.

Two pruning sessions were carried out: formation on 21/08/2010 and maintenance on 06/04/2011. During the formation pruning, the aim was to leave the CCN-51 cocoa plant with a conical shape, maintaining a balance between the lateral branches and the central axis of the plant. The thieving shoots or commonly called "suckers" that appeared on the trunk were eliminated.

Maintenance pruning consisted of removing dry and diseased branches or branches whose foliage did not receive adequate sunlight, as well as dead fruit.

3.1.2.5. Initial shading

The initial shading of the cacao tree is a necessity, not only because it is an umbrella plant, but also because it contributes to a longer crop life and more regular economic production. The adaptation of the cacao tree to the semi-arid region involves the need to create a microclimate that allows the plant to develop normally as if it were in its natural habitat, which is why a consortium between cacao and maize (Zea mays) was set up to provide shade at the beginning of the cacao tree's growth (Figure 11A). This was done on two occasions, on 10/11/2009 and 22/04/2010.

3.1.2.6. Windbreaker

The need for a windbreak was verified by the dehydration of the cocoa leaves and the fact that some plants in the experimental area were close to falling over. On 29/07/2011, banana trees (Musa spp) of the silver variety, Figure 11 B, were planted around the experiment to reduce the influence of the wind on the cocoa plantation.

Experimental design

The experimental statistical design was in randomized blocks in a 4x4 factorial, resulting in 16 treatments made up of water laminas (L) in (mm) and nitrogen doses (N) in kg ha-1, forming the ordered pair (L; N) (Table 4). The treatments were distributed in 4 blocks; each block had 64 plants, 4 plants per plot, with a total of 256 CCN-51 clonal cacao plants in the useful area and 96 plants of the outsider variety "cacao comum" as a border. The treatments were randomly distributed in the experimental area using a sampling technique using numbered papers.

Table 4: Treatments used during the field experiment. Jequié- BA, 2012.

TREATMENTS	From 09/08 /2010 until 19/06/2011		From 09/08 /2010 to 30/01/2012	
	WATER(L)	DOSE (N)	WATER(L)	DOSE (N)
L1N1	1146,35	249,3	1384,52	318,3
L2N 1	1335,66	307,2	1653,219	318,3
L3N1	1525,578	365	1922,521	318,3
L4N1	1717,576	422,9	2193,908	318,3
L1N2	1146,35	249,3	1384,52	405,8
L2N2	1335,66	307,2	1653,219	405,8

L3N2	1525,578	365	1922,521	405,8
L4N2	1717,576	422,9	2193,908	405,8
L1N3	1146,35	249,3	1384,52	493,1
L2N3	1335,66	307,2	1653,219	493,1
L3N3	1525,578	365	1922,521	493,1
L4N3	1717,576	422,9	2193,908	493,1
L1N4	1146,35	249,3	1384,52	580,6
L2N4	1335,66	307,2	1653,219	580,6
L3N4	1525,578	365	1922,521	580,6
L4N4	1717,576	422,9	2193,908	580,6

3.1.2.7. Irrigation and Fertigation Systems

The 65m-high Pedra do Rio de Contas dam has a water volume of 300,000.00 m^3 and a drain with a flow of 5.000.00 m^3 s^{-1} drained water to the booster pump, which belongs to the Fazenda Velha irrigation district, installed near the experimental area, which distributed the water in a pressurized manner to the rural properties in the region, This water supplied the localized drip irrigation system in the experimental area, made up of 2 drip strips per plant row, each 41 m long, with Streamline self-compensating drippers, spaced 0.70 m apart, working at an average flow rate of 1.6 L h^{-1} and a pressure of 196 kPa.

For irrigation management, 16 Rain Bird solenoid valves were used, which were controlled by a 6-sector irrigation multi-scheduler according to the sketch of the experimental area (Figures 12 A and 13).

The system consisted of 5 trestles, the first located at the beginning of the area with the following materials: suction cup, pressure reducer, glycerine manometer to measure the pressure in kg $_{cm-2}$ and a disk filter, and the 4 trestles along the main line to carry out the fertigation, consisting of the following materials: registers, venturi, suction cups, disk filters, PVC handles and containers to prepare the solution to be applied (Figure 12 B).

The water distribution system was assembled using 25 mm, 32 mm and 100 mm Tigre PVC pipes; 18 "Y" connectors for adapting the drip strips to the main pipe, ring connector, knees, bends, hoses and other accessories.

Figure 12. Automated localized irrigation system (A); main trough on the left and N3 fertigation trough on the right (B). Jequié BA, 2012.

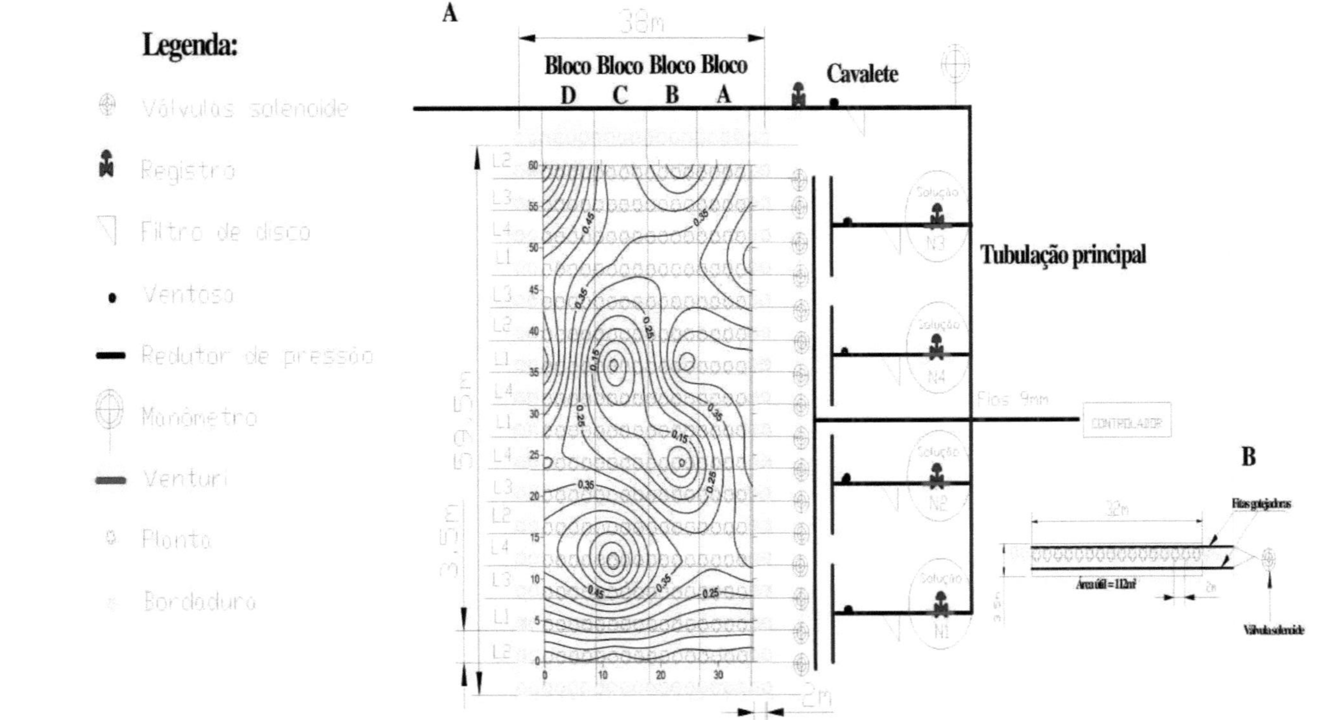

Figure 13. A- Sketch of the experimental area; B-Detail of an irrigation line. Jequié-BA, 2012.

3.1.2.8. Fertilizer management
3.1.2.8.1. Planting

In the 0.40 x 0.40 x 0.40 m pits (Figure 11C), 70 g (pit)$^{-1}$ of FTE BR12, a chemical compound made up of micronutrients, was applied to 140 g (pit)$^{-1}$ of MAP (52% P2O5 + 11% N), i.e. 72,8 g (pit)$^{-1}$ of phosphorus pentoxide (P_2O_5) plus 15.4 g (pit)$^{-1}$ of nitrogen (N_2), adding 3 liters of goat manure to this mixture as an organic fertilizer.

3.1.2.8.2. Plant formation

Throughout the experiment, 194.3 kg ha^{-1} of potassium K_2O were applied, using potassium sulphate (50%) as the source, applied via irrigation water; each application was 2 g (plant)$^{-1}$ and the sulphate was not differentiated, i.e. all the plants received the same amount of potassium.

With regard to nitrogen, the most important nutrient in cocoa tree nutrition, urea (50% N) was used and applied via irrigation water on a weekly basis, taking 3 g N (plant) as a standard^{-1} . This was considered to be the N2-100% dose. The percentages of N1-70, N3-130 and N4-160 % were based on this standard dose. The four nitrogen doses were based on the values indicated for cocoa growing in the first three years according to CEPEC/CEPLAC (2009).

The total doses used throughout the experiment up to the date of the last evaluation of cocoa growth parameters on 16/06/2011 were: N1= 249.30 kg ha^{-1} ; N2 = 307.20 kg ha^{-1} ; N3 = 365 kg ha^{-1} ; N4 = 422.9 kg ha^{-1} .

3.1.2.8.3. Production

Harvesting of the cocoa fruit began on June 20, 2011 and ended on January 29, 2012, when the cocoa tree was two years and two months old; these harvests were carried out sporadically when the cocoa fruit was at the point of ripeness. The total amount of nitrogen applied was N1 = 318.3 kg ha^{-1} ; N2 = 405.8 kg ha^{-1} ; N3 = 493.1 kg ha^{-1} ; N4 = 580.6 kg ha^{-1} (Table 5).

The same proportions were kept as those applied during the crop's formation or growth period, as the cocoa tree would probably only enter the productivity period after the third year of planting.

3.1.2.9. Irrigation management

In the field, four irrigation rates were applied daily whenever there was no rainfall; these were based on the gross rate (GL) needed to replenish the evapotranspirated water: 0.60 GL, 0.80 GL, 1 GL and 1.2 GL. The daily readings at 7:00 hours evaporation data from the Class "A" tank installed on the rural property and the tank coefficient (Kp), which according to Doorembos and Pruitt (1977), for an average wind speed of 4 m s^{-1} (345.6 Km h^{-1}) and relative humidity of 64% with a 1 m border radius is equal to 0.6, were used to estimate the reference evapotranspiration ETo and from this data calculate the net irrigation slurry using Equation 3.

Up until the end of the experiment on 29/01/2012, when the cocoa trees had been planted for two years and two months, the total values of gross water laminae that were applied considering Kc = 0.6; KL = 0.76 according to Equation 6 and UE = 0.9 according to Equation 9, were L1 = 1384.52 mm, L2 = 1653.22 mm, L3 = 1922.52 mm and L4 = 2193.91 mm. These values do not accumulate the actual rainfall that occurred during the experiment (Table 6).

Table 5. Doses of nitrogen (g plant^{-1}) applied to the plots in the experimental area. Jequié-BA, 2012.

MONTHS	1st year in the field 11/2009 to 10/2010				2nd year in the field 11/2010 to 10/2011				3rd year in the field 11/2011 to 01/2012			
	N1(g)	N2(g)	N3(g)	N4(g)	N1(g)	N2(g)	N3(g)	N4(g)	N1(g)	N2(g)	N3(g)	N4(g)
NOVEMBER					8,4	12	15,6	19,2	8,4	12	15,6	19,2
DECEMBER	10	10	10	10	8,4	12	15,6	19,2	8,4	12	15,6	19,2
JANUARY	10	10	10	10	8,4	12	15,6	19,2	3,4	7	10,6	14,2
FEBRUARY	10	10	10	10	8,4	12	15,6	19,2				
MARCO	10	10	10	10	8,4	12	15,6	19,2				
APRIL	10	10	10	10	8,4	12	15,6	19,2				
MAY	10	10	10	10	8,4	12	15,6	19,2				
JUNE	10	10	10	10	8,4	12	15,6	19,2				
JULY	10	10	10	10	8,4	12	15,6	19,2				
AUGUST	6,3	6	7,8	9,6	8,4	12	15,6	19,2				
SEPTEMBER	8,4	12	15,6	19,2	8,4	12	15,6	19,2				
OCTOBER	8,4	12	15,6	19,2	8,4	12	15,6	19,2				
TOTALS	103,1	112	119	128	100,8	144	187,2	230,4	20,2	31	41,8	52,6

Table 6. Gross water depths (mm) applied to the plots in the experimental area. Jequié-BA, 2012.

MONTHS	1st year in the field 11/2009 to 10/2010				2nd year in the field 11/2010 to 10/2011				3rd year in the field 11/2011 to 01/2012			
	L1(mm)	L2(mm)	L3(mm)	L4(mm)	L1(mm)	L2(mm)	L3(mm)	L4(mm)	L1(mm)	L2(mm)	L3(mm)	L4(mm)
NOVEMBER	14	14	14	14	65,86	84,48	103,1	121,72	35,64	50,9	66,09	81,64
DECEMBER	85,0	85,0	85,0	85,0	52,99	70,56	88,13	105,77	39,41	52,57	65,73	78,89
JANUARY	85,0	85,0	85,0	85,0	67,06	89,39	111,79	124,12	51,82	61,78	79,72	91,85
FEBRUARY	85,0	85,0	85,0	85,0	69,51	92,68	110,85	129,02				
MARCO	85,0	85,0	85,0	85,0	76,86	102,48	128,1	133,3				
APRIL	50	50	50	50	52,36	69,79	87,29	104,72				
MAY	40	40	40	40	31,64	42,21	52,71	63,28				
JUNE	40	40	40	40	32,06	42,77	53,48	64,19				
JULY	40	40	40	40	34,79	46,34	54,96	69,51				
AUGUST	33,8	44,24	55,23	68,32	32,55	43,4	54,18	65,03				
SEPTEMBER	40,18	53,55	66,99	80,36	40,74	54,39	67,97	81,55				
OCTOBER	49,14	55,52	71,97	98,35	54,11	72,17	90,23	108,29				
TOTALS	647,12	677,31	718,19	771,03	610,53	810,66	992,79	1210,5	126,87	165,25	211,54	262,09

$$LL = ECA_{média}.K_p.K_c.K_l \qquad (3)$$

Where:

LL = liquid lamina, mm;

ECA_m average= average evaporation in the class A tank over the week;

K_p = tank coefficient, dimensionless;

K_c = crop coefficient, dimensionless;

K_L = adjustment factor due to localized application, which is given by equation 4 ;

To calculate the net irrigation water used the cocoa crop coefficients (Kc) which, according to FAO bulletin 56, are: initial phase 1, intermediate phase 1.05 and final phase 1.05.

Daily evaporation calculations were made by taking the difference between two consecutive evaporation readings in mm, plus the effective rainfall in mm, Equation 4.

$$ECA_{média} = (Leitura\ 2 - Leitura\ 1) + P_e \qquad (4)$$

Where:

P_e = effective precipitation.

According to EMBRAPA (2009), the effective rainfall (P_e), which reaches the soil, can be deduced from the total rainfall (PT) as follows: The actual capacity of water in the soil available to the plants is calculated CRA (Equation 4), which will be taken as the limit value for P_e, i.e. if $P_t >$ CRA then the effective rainfall is $P_e = CRA$ If $P_t < CP/$ then $P_e = P_t$.

$$CRA = (CC - PM).\ da.Z.f.10^{-1} \qquad (5)$$

Where:

CC = field capacity, (%);

PM = permanent soft spot, (%);

da = bulk density of the soil, g. cm" $;^3$

= = depth of the root system, cm;

f = plant water availability factor, dimensionless.

P_t = total rainfall, mm.

The values for field capacity, permanent muck point and bulk density were taken from Table 1; the depth of the root system for the cocoa crop considered by the region's technicians is 0.4 m and the water availability factor used was 0.5.

For the localized irrigation system, the fact that only part of the area is being irrigated was taken into account, and a correction factor K_t was used to calculate the net water lamina. This value, according to Feveres (1981) for fruit trees, is calculated when the percentage of wet or shaded area P_w is in the range 20% $< Pw <$ 65%, using Equation 6.

$$K_l = 1,09.\ Pw.\ 10^{-2} + 0,30 \qquad (6)$$

Where:

K = factor due to localized application, dimensionless;

P_w = percentage of wet or shaded area (%).

Field tests were carried out to check the diameter of the bulb wetted by the emitter in order to calculate the percentage of wetted area, finding an average value of 0.78 m, which, according to Equation 7, for two drip strips per plant row, gave a percentage of wetted area of 42%, a value within the standards accepted by most bibliographies, which quote values above 33% for semi-arid regions (Bernardo et al., 2007), using 6 emitters per plant. -

$$Se \geq Se' \qquad P_w = \frac{NEP.\ Se'.(Se'+W)}{2.\ S_n.\ S_f}.100 \qquad (7)$$

Where:

Se = spacing between emitters, m;

Se' = spacing between emitters to have a continuous wetted soil volume, m;

NEP = number of emitters per plant;

$T"$ = maximum diameter of the bulb wetted by the emitter, m;

S_p = spacing between plants, m;
S_f = spacing between plant rows, m.

$$Se' = 0,8.W \qquad (8)$$

The calculation of the gross blade used the coefficient of uniformity of emission (CUE) of water, whose value found in the field using the methodology of Kelly and Karmelly (1974) was 90%, and the leachate ratio calculated from the water analysis (Table 2), whose value was less than 0.1 and was therefore negligible.

To calculate the uniformity of water emission (CUE) in the field (Equation 9), the flow rate was obtained at four points along the lateral line, at the first dripper, at the drippers located 1/3 and 2/3 of the way along and the last dripper. The lateral lines selected along the drip line were the first, those at 1/3 and 2/3 of the length and the last lateral line (Figure 14).

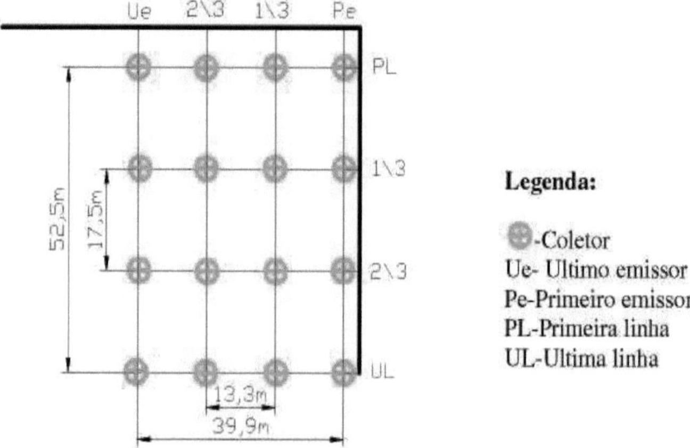

Figure 14. Distribution of the collectors in the field for the water distribution uniformity test. Jequié-BA, 2012.

To determine the emission uniformity coefficient, three tests were carried out to collect the volume of water in $L.s^{-1}$ in a fixed time of 10 minutes, in order to calculate the flow rate at each selected emitter. The tests took place at 8:00 am with a pressure of 196 kPa at the manometer located in the main trestle.

$$CUE = 100 . \left[\frac{q_n}{\bar{q}}\right] \qquad (9)$$

Where:
q_n = average of 25% of the lowest flows;
q = average of all the flows collected.

To calculate the leaching ratio (Equation 10), the electrical conductivity of the irrigation water $CEa = 0.24$ dS m^{-1} was used, according to Table 2, and the estimated electrical conductivity for cocoa of 2 dS m^{-1}, with the gross irrigation laminas used being sufficient to leach the salts.

$$R_L = \frac{CEa}{2.CE_c} \qquad (10)$$

Where:

R_L = leaching rate, dimensionless
ECa = electrical conductivity of irrigation water. $dS.$ m^{-1} .
EC_c = estimated electrical conductivity for cocoa, dS. m^{-1}

Equation 11 will be used to calculate the application efficiency.

$$E_a = 0,9 . UE \qquad (11)$$

The calculation of the gross irrigation rate, which was used as a reference to replace the evapotranspired water and from which the other water rates that made up the treatments applied in the field were referenced, was carried out using Equation 12 described by Bernardo et al. (2006).

$$LB = \frac{LL}{CUE. \ (1-R_L)} \qquad (12)$$

Where:
LB = gross lamina, mm; .
To calculate the volume of water applied per plant (VP), we will use Equation 13.

$$VP = LB. S_p. S_f \qquad (13)$$

Where:
VP= volume of water applied per plant, L .
Six emitters were used per plant, with an average flow rate of 1.6 L h^{-1} for each emitter, in order to calculate the application time of the gross irrigation blade in mm. The application times of the irrigation blades were calculated from Equation 14.

$$Ta = \frac{Vp}{NEP. \ q_{méd}} \qquad (14)$$

Where:
Ta= application time, h;
$q_{méd}$ = average emitter flow rate, Lh^{-1} .

3.1.2.10. Fertigation Management

Fertilization of the cocoa trees was carried out in accordance with the principles of fertigation, with an initial application of 25% of the time allocated to applying the gross irrigation rate, in order to hydraulically balance the irrigation units; Urea was then injected as a nitrogen source, applied in the appropriate proportions according to the dose established as the weekly application standard of 3 g N (plant)$^{-1}$ and considered as N2 (100%), and the other levels of N1 (70%), N3 (130%) and N4 (160%) in relation to the standard dose. Potassium sulphate was used as the potassium source, and the same amount of 2 g K (plant)$^{-1}$ was applied weekly to all the plants.

These mixtures were diluted in a ten-liter container with a discharge time of five minutes. The system injected the fertilizer into the pipes through the four venturis after filtering. The final 25% of the gross irrigation time was used to completely flush the irrigation system and carry the fertilizers from the surface to the deeper layers of the soil.

Fertigation was carried out weekly on Saturdays at 8 a.m. in the four venturis, each with a suction flow of 2 L (min)$^{-1}$ and a pressure of 196 kPa.

3.1.2.11. Analysis of growth and production characteristics

Figure 15 shows the dates of the evaluations that were carried out on the growth and production characteristics of the CCN-51 clonal cacao tree, in a natural environment.

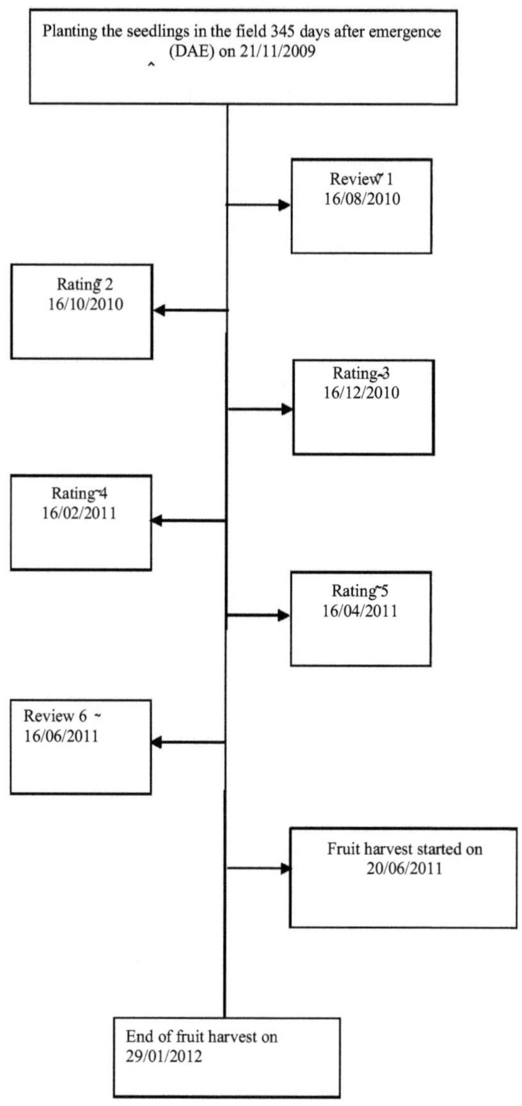

Figure 15. Organizational chart of the assessments carried out in the natural environment. Jequié-BA, 2012

3.1.2.11.1. Growth characteristics

During the experimental period from 21/11/2009 to 16/06/2011, the plant's growth characteristics were measured over the days after transplanting to the field (DAT), height (AP) in cm and stem diameter (DC) in mm on six occasions: (265DAT), (325DAT), (385DAT), (445DAT), (505DAT) and (565DAT) as shown in Figure 15, with height measured from ground level to the basal part of the last leaf with a 3 m tape measure and stem diameter at ground level with a 150 mm caliper accurate to 0.05 mm. The absolute and relative growth rates of the characteristics evaluated were found.

3.1.2.12. 2. Production characteristics

The following plant production characteristics were determined from June 20, 2011 to February 8, 2012: Number of Fruits per Plant (NFP), Fruit Length (CF), Fruit Diameter (DF), Fruit Weight (PF), Number of Seeds per Fruit (NSF), Kernel Weight per Fruit (PAF) and Production (P), The length of the fruit was measured with a 1 m tape measure and the diameter of the fruit with a 150 mm steel caliper to an accuracy of 0.05 mm, the weight of the fruit and kernels with a precision scale to 4 decimal places and the others by direct counting.

In order to collect data on the kernel weight variable, the seeds had to be exposed to full sunlight for a period of approximately three days until they were at the point of sale and then weighed. The other characteristics were measured as soon as the fruit was removed from the cocoa tree.

Figure 16. Identification of treatments on cocoa fruits. Jequié-BA, 2012.

3.1.2.13. Statistical and mathematical analysis

The data was subjected to analysis of variance and regression using ASSISTAT (2011) and SAS software. The response surfaces were constructed using the STATISTIC 7.0 program, and the relative maximum points were determined using the partial derivatives of the factors included in the model obtained from the analysis of the significance of the coefficients, as well as calculations relating to the relative growth of the plants and fruit.

Table 7. Simplified model for analysis of variance and regression of the parameters that were evaluated. Jequié-BA, 2012.

Source of variation	GL	F-statistics			
		AP (cm)	DC (mm)	NFP TF DF (cm)	PF (Kg)
Blade (L)	I - 1				

Linear Regression	1				
Quadratic Regression	1				
Cubic Regression	1				
Nitrogen (N)	$J - 1$				
Linear Regression	1				
Quadratic Regression	1				
Cubic Regression	1				
Interaction (L x N)	$(I - 1).(J - 1)$				
Block	$K - 1$				
Waste	$(IJ - 1).(K - 1)$				

CV(%)

Using the cocoa crop production index for the first year of productivity, the points of relative maximum physical and economic productivity were determined.

3.1.2.14. Absolute and relative growth rate.

The absolute growth rate is defined as the increase between two samples throughout the cycle, indicating the plant's growth rate. The relative growth rate is directly related to the size reached in the previous period and is therefore a function of the initial size at the beginning of the observation period. Equations 15 and 16 were used to calculate the absolute and relative rates for cocoa tree height and diameter.

$$TCR = \frac{ln C_F - ln C_I}{T_F - T_I} \qquad (15)$$

$$TCA = \frac{C_F - C_I}{T_F - T_I} \qquad (16)$$

Where:

TCR = relative growth rate;

TCA= absolute growth rate;

ln C_F = Neperian logarithm of the final length in (cm);

ln C_I = Neperian logarithm of the initial length in (cm);

C_F = final length in (cm);

C_I = initial length in (cm);

T_F =Final time in days;

T_I =Initial time in days.

3.1.2.15. Production function

The production function was found from the regression analysis between the dependent variable and the independent variables, whose second degree polynomial model is described by several authors (Frizzone, 1993). However, other regression models such as the cubic model have been much better suited to the data to describe the yields of various crops. Using cocoa bean yield values, water table and nitrogen doses, three regression models indicated by Equations 17, 18 and 19 for two independent variables were analyzed using the significance of the coefficients and the coefficient of determination (R^2).

A fourth model was proposed (Equation 20), as the previous models were not significantly representative for the variables number of seeds per plant and kernel weight per fruit.

The cocoa bean production function estimated for two independent variables was represented by the regression model described in Equation 18 .

$$\hat{Y} = K_0 + K_1 N + K_2 N^2 + K_3 N^3 + K_4 L + K_5 L^2 + K_6 L^3 + K_7 L.N \qquad (17)$$

$$\hat{Y} = K_0 + K_1 N + K_2 N^2 + K_3 N^3 + K_4 L + K_5 L^2 + K_6 L.N \qquad (18)$$

$$\hat{Y} = K_0 + K_1 N + K_2 N^2 + K_3 L + K_4 L^2 + K_5 L.N \qquad (19)$$

$$\hat{Y} = K_0 + K_1 N + K_2 N^2 + K_3 N^3 \qquad (20)$$

Where:

$\hat{Y}$ = estimated maximum cocoa fruit yield *(kg. ha)*;$^{-1}$

L = total water level in *mm;*

N = nitrogen dose in *(kg. ha)*;$^{-1}$

$$\hat{K}_0,\ \hat{K}_1,\ \hat{K}_2,\ \hat{K}_3,\ \hat{K}_4,\ \hat{K}_5,\ \hat{K}_6\ e\ \hat{K}_7 \quad = \text{estimated}$$

coefficients of the regression models analyzed.

For two variable factors, the conditions necessary for the existence of a maximum point or optimization of the production function are described from the partial derivatives of the function $\hat{Y} = f(L, N),$ i.e. a maximum point is considered if the sum of the derivatives and its Hessian determinant is greater than zero Equations 21, 22 and 23.

$$\frac{\partial \hat{Y}}{\partial L} = 0 \qquad \frac{\partial \hat{Y}}{\partial N} = 0 \tag{21}$$

$$\frac{\partial^2 \hat{Y}}{\partial^2 L_{(L,N)}} + \frac{\partial^2 \hat{Y}}{\partial^2 N_{(L,N)}} < 0 \tag{22}$$

$$\begin{vmatrix} \dfrac{\partial^2 \hat{Y}}{\partial^2 L} & \dfrac{\partial^2 \hat{Y}}{\partial L.\partial N} \\ \dfrac{\partial^2 \hat{Y}}{\partial N.\partial L} & \dfrac{\partial^2 \hat{Y}}{\partial^2 N} \end{vmatrix}_{(L,N)} > 0 \tag{23}$$

Once these conditions have been met, the point (L, N) is the maximum point of the production function that was estimated through regression analysis. This maximum value for the water table and nitrogen dose is found by solving the equation system formed by Equations 24 and 25.

$$\frac{\partial \hat{Y}}{\partial L} = K_4 + 2K_5 L + K_6 N = 0 \tag{24}$$

$$\frac{\partial \hat{Y}}{\partial N} = K_1 + 2K_2 N + 3K_3 N^2 + K_6 L = 0 \tag{25}$$

3.1.2.16. Simplified economic analysis of treatments

To calculate net revenue we have

$$R_L = R_B - C_T \tag{26}$$

Where:
R_B = gross revenue in R\$;
C_T = total cost in R\$.

$$R_B = P_Y . \hat{Y} \tag{27}$$

Where:
P_Y = cocoa price in $(R\$.kg^{\sim ly})$;
Where:

$$C_T = C_V + C_F \tag{28}$$

C_V ,=variable cost R\$;
C_P = total fixed cost in $(R\$.\ kg\ )^{.-1}$

$$C_V = P_L . L + P_N . N \tag{29}$$

Where:
P_L = price of the water blade in $(R\$.\ mm^{-1}.\ ha^{-1})$;
P_N = price of nitrogen in $(R\$.\ kg\ )^{.-1}$

Substituting Equations 27, 28 and 29 into Equation 26 we find the mathematical expression that provides the calculation of the net revenue (R_L) in R\$, for the combinations of water blade and nitrogen doses.

$$R_L = P_Y . \hat{Y} - P_L . L - P_N . N - C_F \tag{30}$$

Solving the system of equations formed by Equations 33 and 34 we find the point (L_E, W_E) which will be considered the point of maximum, i.e. the point that maximizes net revenue if it meets the conditions described in maximum of the production function.

$$\frac{\partial R_L}{\partial L} = P_Y \cdot \frac{\partial Y}{\partial L} - P_L \tag{31}$$

$$\frac{\partial R_L}{\partial N} = P_Y \cdot \frac{\partial Y}{\partial N} - P_N \tag{32}$$

$$\frac{\partial Y}{\partial L} = K_4 + 2K_5 L + K_6 N = \frac{P_L}{P_Y} \tag{33}$$

$$\frac{\partial Y}{\partial N} = K_1 + 2K_2 N + 3K_3 N^2 + K_6 L = \frac{P_N}{P_Y} \tag{34}$$

The data used for the economic analysis was obtained as follows: the price of cocoa supplied by producers in the region was R$ 5.07.kg^{-1} ; the price of water according to Frizonne et al. (1996), the price of water can be considered to be equal to the value of the electricity tariff, so the price of a millimeter of water is given by the cost of electricity (R$) divided by the amount of water applied during the period in mm, which was R$ 0.64mm-1 ha. This cost of electricity consumption was provided by the Bahia State Electricity Company (COELBA).

The cost of nitrogen was estimated based on the price of urea, i.e. 0.45 kg of nitrogen per kg of urea, worth R$ 2.44 per kilo.

The fixed cost is the sum of the production costs plus the annual amortization. The production costs for one hectare of cocoa were provided by the Executive Commission of the Cocoa Farming Plan (CEPLAC), Table 8.

In amortization, the fixed costs are paid off in installments over a certain period of the equipment's useful life, plus a certain interest rate.

$$C_F = C_p + A \tag{35}$$

Where:
C_p =cost of crop production in (F$. ha^{-1}) ;
The annual amortization of investments in (F$. ha)$.$^{-1}$

The annual amortization of investments in irrigated cocoa cultivation was estimated based on the principle of capital recovery expressed in Equation 36.
Where:

$$A = I_o \cdot FRC \tag{36}$$

A = annual amortization of investments in the irrigation system and land (F$. ha)$;$^{-1}$
I_o = investment in irrigation system and land (R$) ;
$I-7¿C$ = capital recovery factor.

$$FRC = \frac{J.(1+J)^n}{(1+J)^n - 1} \tag{37}$$

Where:
J= annual interest rate, %;
n = useful life of the equipment, year.

Table 8. Production costs of one hectare of CCN-51 clonal cacao. Jequié-BA, 2012.

COST ITEMS	1st Year			
	Unit	Unit	Quantity	Value

1. SUPPLIES				
Seedlings (grafted)	Ud	1,00	1429	1429,00
Insecticide	L	31,00	2	62,00
Phosphorus - P2O5	Kg	2,18	45	98,10
Potassium - K2O	Kg	1,26	20	25,20
Herbicide	L	12,30	3	36,90
Fungicide	Kg	25,00	2	50,00
SUBTOTAL I	**R$**			**1. 701.20**
2. SERVED				
Clean the area	d/H	20,00	4	80,00
Building a nursery	d/H	20,00	4	80,00

Filling and planting bags	d/H	20,00	3	60,00
Cultivation	d/H	20,00	3	60,00
Opening the pits	d/H	20,00	3	60,00
Planting seedlings	d/H	20,00	5	100,00
Rocking	d/H	20,00	8	160,00
Fertilization	d/H	20,00	4	60,00
Pest control	d/H	25,00	2	50,00
Formation pruning	d/H	20,00	1	20,00
Conductive and productive pruning	d/H	20,00	1	20,00
Herbicide application	d/H	20,00	1	20,00
Harvesting	d/H	20,00	2	40,00

SUBTOTAL II	**R$**			**810,00**
TOTAL	**R$**			**2511,20**

To calculate the aniorti/acao estimate, the value of localized irrigation equipment for one hectare with two drip tapes of R$ 4600.00 was taken into account; one hectare of bare land in Fa/enda Velha in Jequié-BA whose estimated value is R$ 2000.00 plus the price of building the motor pump set house of R$ 400.00 supplied by the
Agroconsult irrigation projects. We considered an interest rate of 12% per year and a useful life of 10 years for the equipment, values used in most agricultural negotiations, and we will consider zero as the residual value at the end of the useful life.

3.1.2.17. Water efficiency

Water use efficiency was determined by the ratio between cocoa tree yield, obtained from the yield function and total water levels (L) for each treatment (Equation 38) according to Doorembos and Kassan (1994):

$$EUA = \frac{P_y}{L} \tag{38}$$

Where:
EUA= water use efficiency in kg ha^{-1} mm^{-1} ;
Py = cocoa productivity in kg ha^{-1} ;
L= the applied blade in mm.

3.1.2.18. Nitrogen use efficiency (NUE)

Nitrogen use efficiency is defined as the number of kilos added to cocoa production for each kilo of nitrogen added to the fertilizer.

$$EUN = \frac{P_{tr}-P_{te}}{N_t} \tag{39}$$

Where:
EUN - nitrogen use efficiency for the cocoa crop, dimensionless;
P_{tr} - productivity in treatment "t" in kg ha^{-1} ;
P_{te} - productivity in the control treatment kg.ha^{-1} ;
N_t - amount of nitrogen applied in treatment "t" in kg.ha^{-1}
Since P_{te} doesn't exist, we have

$$EUN = \frac{P_{tr}}{N_t} \tag{40}$$

Moll et al. (1982) quote a similar mathematical expression for calculating EUN (Equation 41):

$$EUN = \frac{G_w}{N_s} \tag{41}$$

Where:
Gw = mass of grains, kg;
Ns = mass of N applied to the soil, kg ha^{-1} .

Chapter 4

4. RESULTS AND DISCUSSION

4.1. Protected environment

Tables 9, 10 and 11 show the growth in height, diameter and number of leaves of the CCN-51 cocoa seedlings.

The seedlings showed an average height of 30.51 cm at 195 days after emergence (DAE) and 43.46 cm at 345 days after emergence, a relative growth of 42.44 % (Table 9). Adu-Ampomah et al. (2003) working with crosses of cloned cocoa seedlings at 6 months of age found average heights ranging from 48.6 to 55 cm. Efron et al. (2003) working with clonal cocoa seedlings originating from rootstock found heights of 32.9 cm, 46.9 cm and 54 cm at 3, 6 and 9 months after sowing, respectively.

Table 9. Average height of CCN-51 cocoa seedlings. Jequié-BA, 2012.

DAE	Height (cm)			
	Average	CV(%)	Min	Max
195	30,51	3,17	29,1	32,2
269	37,83	5,19	34,10	40,60
345	43,46	5,70	38,10	47,40

The stem diameter values of the seedlings evaluated at 195, 269 and 345 DAE were 7.1, 8.2 and 9.1 mm, respectively. They showed a relative growth of 15.59 % and 10.21 % from 195 to 269 DAE and from 269 to 345 DAE, respectively (Table 10). Frazao et al. (1984), working with common cocoa seedlings in a greenhouse, found average diameters of 1.7 and 5 mm at 30 and 165 days of age, showing a relative growth of 96%. These same authors also identified linear growth for this growth characteristic.

Efron et al. (2003), working with clonal cocoa seedlings originating from scion rootstocks, found stem diameter values of 3, 6.3 and 10.1 mm, respectively, at 3, 6 and 9 months after sowing. Souza Junior and Carmello (2008), analyzing cocoa seedlings cloned by cuttings at 145 days old, found the average stem diameter of the seedlings to be 3.92 mm.

Table 10. Average diameter of CCN-51 cocoa seedlings. Jequié-BA, 2012.

DAE	Diameter (mm)			
	Average	CV(%)	Min	Max
195	7,12	2,00	3,80	7,40
269	8,23	4,09	7,80	9,00
345	9,07	4,46	8,5	9,9

Table 11. Average number of leaves of CCN-51 cocoa seedlings. Jequié-BA, 2012.

DAE	Number of leaves			
	Average	CV(%)	Min	Max
195	8,88	6,31	8,10	10,30
269	12,14	9,55	10,7	14,8
345	17,04	3,79	14,1	21,6

It can be seen that for the three periods when the seedlings were evaluated at 195, 269 and 345 days after emergence, an average of 8.9, 12.1 and 17.0 leaves were found per plant respectively (Table 11). Throughout the experimental period, the growth in the average number of leaves at 269 days compared to 195 days and at 345 days compared to 269 days was 36.71% and 40.36%, respectively.

Souza et al. (2008), working with various cocoa clones, including the CCN-51 clone, irrigated, with a substrate composed of a mixture of soil: sand: organic matter in a 3:1 ratio, found average values for the number of leaves of 12 units at 150 days of age and 15 leaves at 270 days of age; these averages showed significant differences in relation to the averages of the PH-16 and PS-1319 clones.Similarly, Santos and Ribeiro (2008), researching cloned cocoa seedlings under irrigation and with a substrate prepared with clay, sand and goat manure in a 3:1:1 ratio, found that the average number of leaves for the CCN-51 clone was 15 at 270 days of growth, immediately lower than that of the PH-16 and PS-1319 seedlings.

Almeida and Chaves (2010) found linear growth for the parameters number of leaves, diameter and plant height of CCN-51 cocoa seedlings and an increase of 93%, 27% and 24% at 336 days compared to 234 days of age.

4.2. Natural Environment.

4.2.1. Plant growth

The water levels (L) applied in this study significantly affected plant height, with probability $(0.01 < p < 0.05)$ in evaluation 6, while in the other evaluations these levels showed no influence on this growth characteristic, as shown in Table 12. These water levels influenced the growth of the stem diameter in all the evaluations carried out, with the exception of evaluation 4, which showed a non-significant result with probability $(0.01 < p < 0.05)$ and $(p < 0.01)$, according to Table 13.

An increase in nitrogen (N) levels led to a significant increase $(p < 0.01)$ in the growth characteristics evaluated: plant height and stem diameter, which were significantly affected at the levels studied for all the evaluations carried out according to Tables 12 and 13. The results of the interaction between L and N did not significantly affect the characteristics analyzed (Table 12 and 13).

The regression models that best represented the evolution of plant height and stem diameter as a function of the water levels in the 6 evaluations carried out were linear and for the nitrogen doses they were quadratic (Table 12 and 13).

Table 12. Summary of the analysis of variance and regression of the variable plant height (PH) in (cm) in 6 evaluations, submitted to water laminas and nitrogen doses. Jequié BA, 2012.

Source of Variation	GL			Court	Middle		
		AP1	AP2	AP3	AP4	AP5	AP6
Block	3	$16,88^{ns}$	$33,83^{ns}$	$56,89^{ns}$	$21,61^{ns}$	$55,20^{ns}$	$87,10^{ns}$
L	3	$48,27^{ns}$	$91,38^{ns}$	$139,17^{ns}$	$218,30^{ns}$	$214,85^{ns}$	$312,24^{*}$
N	3	$321,24^{**}$	$319,88^{**}$	$433,70^{**}$	$558,16^{**}$	$719,56^{**}$	$980,57^{**}$
LxN	9	$27,05^{ns}$	$43,81^{ns}$	$73,62^{ns}$	$142,13^{ns}$	$131,78^{ns}$	$139,66^{ns}$

Waste	45	30,43	40,32	52,91	84,60	91,19	79,74
CV (%)		7,01	6,96	6,38	6,92	6,21	5,5
Blades (L)							
Linear Regression	1	$35,53^{ns}$	$62,38^{**}$	$80,25^{*}$	$111,66^{*}$	$127,20^{*}$	$211,79^{**}$
Quadratic Regression	1	$0,55^{ns}$	$3,73^{ns}$	$21,00^{ns}$	$12,82^{ns}$	$28,56^{ns}$	$20,63^{ns}$
Cubic Regression	1	$0,11^{ns}$	$2,41^{ns}$	$3,12^{ns}$	$39,25^{ns}$	$5,38^{ns}$	$6,26^{ns}$
Levels Nitrogen (N)							
Linear Regression	1	$91,16^{*}$	$33,18^{ns}$	$28,55^{ns}$	$18,94^{ns}$	$172,58^{ns}$	$358,07^{*}$
Quadratic Regression	1	$118,61^{**}$	$187,78^{**}$	$255,33^{*}$	$372,45^{*}$	$362,98^{*}$	$372,17^{**}$
Cubic Regression	1	$31,17^{ns}$	$18,95^{ns}$	$41,39^{ns}$	$27,24^{ns}$	$4,11^{ns}$	$5,19^{ns}$
Blades (L) in mm			Mé	days			
L1 (1384,52)		80,44	93,56	115,72	134,80	156,00	165,84
L2(1653,22)		79,63	92,07	116,80	137,03	157,19	166,00
L3 (1922,52)		78,07	91,34	113,61	130,47	153,11	161,07

				Mé	days		
L4 (2193,91)		76,52	87,92	111,06	129,11	148,95	156,66
Nitrogen (N) levels in kg.ha^{-1}				Mé	days		
N1(318,30)		78,80	93,20	116,98	136,80	154,40	160,62
N2 (405,80)		73,00	85,69	107,31	125,79	146,90	156,22
N3 (493,10)		78,88	89,90	112,82	130,27	151,20	158,93
N4 (580,60)		83,96	96,09	119,13	138,55	162,76	173,82

Ns; * ; **; non-significance and significance at the 5% (0.01< p< 0.05) and 1% (p < 0.01) probability levels, respectively.

Table 13. Summary of the analysis of variance and regression of the variable Stem Diameter (SD) in (mm) in 6 evaluations, submitted to water tables and nitrogen doses. Jequié BA, 2012.

Source of Variation	GL	Mean Square					
		DC1	DC2	DC3	DC4	DC5	DC6
Block	3	1,70^{ns}	0,87^{ns}	3,05^{ns}	0,12^{ns}	2,73^{ns}	6,60^{ns}
L	3	9,35*	16,34**	13,46*	36,31^{ns}	40,85**	40,81*
N	3	15,44**	16,42**	12,95*	87,39**	45,42**	78,22**

LxN	9	$1,81^{ns}$	$4,04^{ns}$	$6,00^{ns}$	$23,77^{ns}$	$7,72^{ns}$	$18,69^{ns}$
Waste	45	2,68	3,40	4,36	19,10	8,83	11,92
CV (%) Blades (L)		7,68	7,46	6,73	11,19	6,53	6,63
Linear Regression	1	$6,32^{**}$	$12,20^{**}$	$9,74^{**}$	$16,21^{ns}$	$28,40^{**}$	$29,05^{**}$
Quadratic Regression	1	$0,53^{ns}$	$0,002^{ns}$	$0,36^{ns}$	$6,31^{ns}$	$2,14^{ns}$	$1,28^{ns}$
Cubic Regression Levels Nitrogen (N)	1	$0,16^{ns}$	$0,05^{ns}$	$0,0004^{ns}$	$4,71^{ns}$	$0,09^{ns}$	$0,28^{ns}$
Linear Regression	1	$3,79^{*}$	$0,72^{ns}$	$2,72^{ns}$	$34,47^{**}$	$20,11^{*}$	$39,03^{*}$
Quadratic Regression	1	$7,74^{**}$	$11,45^{*}$	$6,80^{ns}$	$25,56^{**}$	$12,80^{ns}$	$13,51^{ns}$
Cubic Regression	1	$0,038^{ns}$	$0,14^{ns}$	$0,003^{ns}$	$5,51^{ns}$	$1,15^{ns}$	$6,12^{ns}$
Blades (L) in mm				Average			
L1 (1384,52)		22,03	25,93	31,90	40,00	46,86	53,55

L2(1653,22)		21,65	25,03	31,51	39,39	46,54	53,14
L3 (1922,52)		21,36	24,40	30,80	39,95	45,14	51,58
L4 (2193,91)		20,25	23,53	29,81	36,82	43,35	50,05
Nitrogen (N) levels in kg.ha^{-1}				Average			
N1(318,30)		21,39	25,24	31,12	38,07	44,74	50,63
N2 (405,80)		20,34	23,91	30,14	37,91	44,44	51,29
N3 (493,10)		20,91	23,84	30,54	37,65	44,72	51,03
N4 (580,60)		22,65	25,90	32,21	42,54	47,99	55,37

Ns; * ; **; non-significance and significance at the 5% (0.01< p< 0.05) and 1% (p < 0.01) probability levels, respectively.

  Plant height and stem diameter were significantly affected by water levels described by a quadratic regression (Figure 17A and 17B). Abdoellah and Notoradiningrat (1993) also found a linear regression to be significant for cocoa stem diameter growth when analyzing the influence of Al / K + Ca + Mg on plant growth.

  However, the plant height values as a function of water levels ranged from 165.86 cm to 156.66 cm (Figure 17 A), i.e. increasing the amount of irrigation water from 1146.35 millimeters to 1717.58 millimeters reduced plant height by 5.5%. In the same way, the increase in irrigation water reduced the stem diameter (Figure 17B) by 6.54%, considering the highest and lowest amounts of irrigation water applied.

  Irrigated cocoa planting in the semi-arid region is an innovation, as there is still no precise information on irrigation and production management. The effect of irrigation on cocoa production depends mainly on the amount and distribution of rainfall.

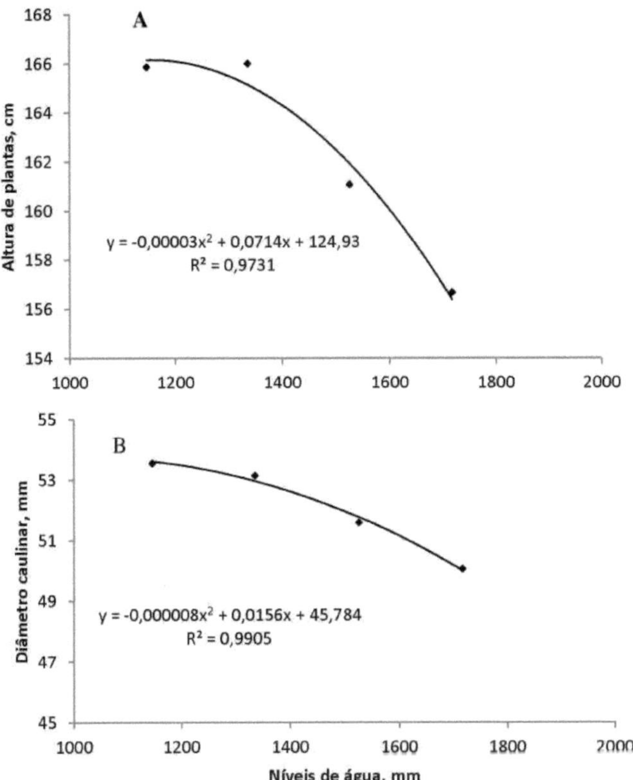

Figure 17 Plant height (A) and stem diameter (B) in relation to water levels.
Jequié-BA, 2012.

According to Purdy and Schmidt (1996), the height of cacao can vary from 5 to 8 m with rainfall ranging from 1250 to 2800 millimeters per year. PROAMAZONIA (2004) and Gramacho et al. (1992) cite similar rainfall values for maximum plant growth. Dias and Resende (2001) and Abdul-Karimu et al. (2003) found that the average stem diameter and plant height of a two-year-old cocoa plant in traditional coca growing regions were 32.7 mm and 150 cm, respectively.

According to Braudeau (1970), cocoa is susceptible to the stress caused by a lack of moisture in the soil. However, several authors have concluded that irrigation in Brazil's semi-arid conditions can be complementary, as long as soil moisture levels are ideal for plant growth and development. However, applying too much water to the soil affects the development of cocoa.

A polynomial regression describes the relationship between nitrogen levels and plant height and stem diameter, representing the contribution of nitrogen fertilization to the growth of these characteristics studied in the experimental area (Figures 18 A and 18 B). Nitrogen is the mineral nutrient most demanded by plants and Chepote et al. (2005) agreed that it is the basis of cocoa nutrient fertilization. The results of this study showed that increasing the level of nitrogen increased plant height and stem diameter.

Plant height varied from 160.62 to 173.82 cm (Figure 18 A) and stem diameter varied from 50.63 to 55.37 mm (Figure 18 B) when 249.31 and 422.9 kg ha^{-1} of nitrogen were applied, respectively, approximately 535 days after transplanting into the field. There was an increase in plant height and stem diameter of 8.22% and 9.36%, respectively, compared to the 69.63% increase in nitrogen.

Souza Junior and Carmello (2008), studying the effects of nitrogen fertilization on the production of cocoa seedlings, observed that plant height and stem diameter responded significantly to a quadratic increase in nitrogen. On the other hand, Leite (2006), studying the growth in diameter and average height of cacao for 630 days under irrigation, found a linear increase in plant diameter and height from the third month after planting, with the value of stem diameter and plant height of 35.8 millimeters and 190.6 cm, respectively.

Chepote et al. (2005), studying a 24-month-old cocoa plant and applying a fertilizer of 240 kg ha-1 of the 30-90-30 formula, showed a diameter of 28.5 mm and a height of 181.7 cm.

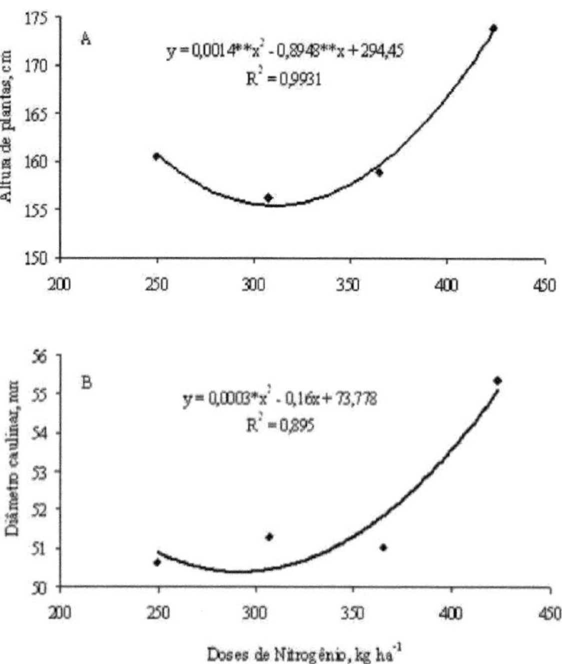

Figure 18 Plant height (A) and stem diameter (B) in relation to nitrogen doses. Jequié-BA, 2012. The sigmoid profiles for variable plant height as a function of days after transplanting (DAT) are shown in Figure 19, corroborating Benincasa (2003) who found a similar profile for this growth characteristic.

The highest absolute and relative growth rates in plant height occurred between 325 and 385 days after transplanting, corresponding to 0.41 cm.day^{-1} and 0.0040 (cm.cm^{-1} .day^{-1}), obtained with 1335.66 mm of water (Table 14). Nitrogen promoted an absolute maximum increase in plant height of 0.40 cm.day^{-1} from 445 to 505 days after transplanting into the field with an application of 422.90 kg.ha^{-1} (Table 14).

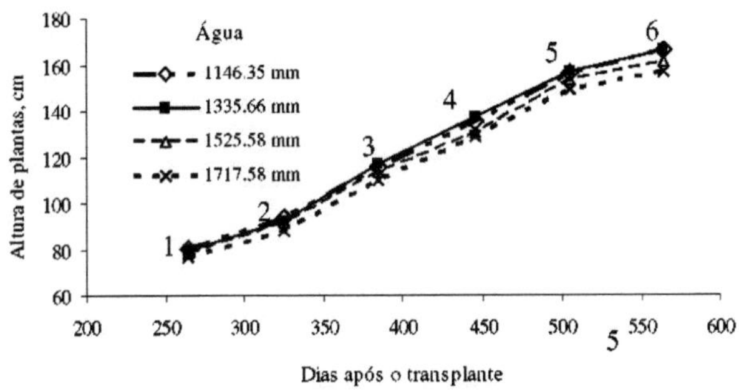

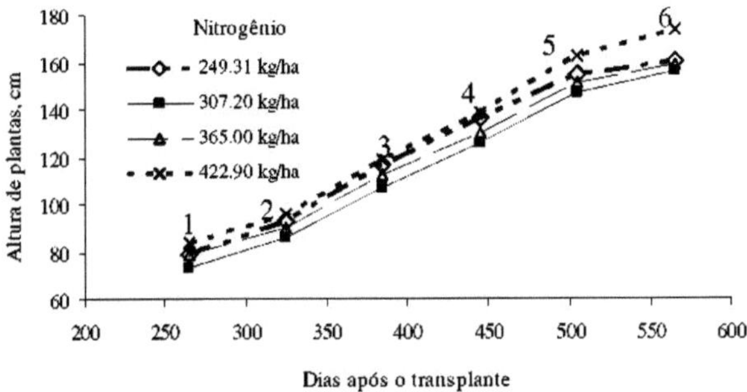

Figure 19 Height of plants subjected to different water and nitrogen treatments as a function of time. Jequié-BA, 2012.

Table 14. Absolute growth rate (ACR) in (cm.day^{-1}) and relative growth rate (RGR) in (cm.cm^{-1} .day^{-1}) of plant height . Jequié-BA, 2012.

Levels -	1---2		2---3		3---4		4---5		5---6	
	TCA	TCR	TCA	TCR	TCA	TCR	TCA	TCR	TCA	TCR
L1	0,22	0,0025	0,37	0,0035	0,32	0,0025	0,35	0,0024	0,16	0,0010
L2	0,21	0,0024	0,41	0,0040	0,34	0,0027	0,34	0,0023	0,15	0,0009

L3	0,22	0,0026	0,37	0,0036	0,28	0,0023	0,38	0,0027	0,13	0,0008
L4	0,19	0,0023	0,37	0,0038	0,32	0,0027	0,33	0,0024	0,13	0,0008
N1	0,24	0,0028	0,04	0,0038	0,33	0,0026	0,29	0,0020	0,10	0,0007
N2	0,21	0,0027	0,36	0,0037	0,31	0,0026	0,35	0,0026	0,16	0,0010
N3	0,18	0,0022	0,38	0,0038	0,29	0,0024	0,35	0,0025	0,13	0,0008
N4	0,20	0,0022	0,38	0,0036	0,32	0,0025	0,40	0,0027	0,18	0,0011

The sigmoid profiles for stem diameter as a function of days after transplanting (DAT) are shown in Figure 20, which are similar to the profiles found for plant height.

The highest absolute growth rate in stem diameter occurred between 385 and 445 days after transplanting, corresponding to 0.17 cm.day^{-1} obtained with 1717.58 mm of water (Table 15). Nitrogen promoted a maximum relative increase in stem diameter of 0.0043 (cm.cm^{-1} .day^{-1}) in the interval from 385 days to 445 days after transplanting with 366.00 kg.ha^{-1} (Table 15).

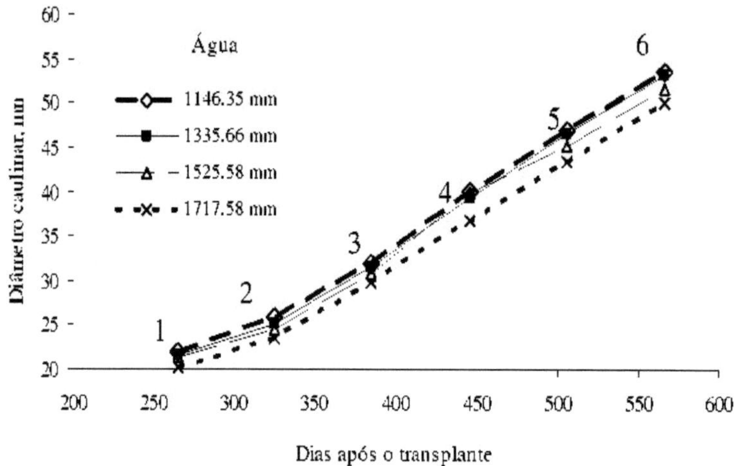

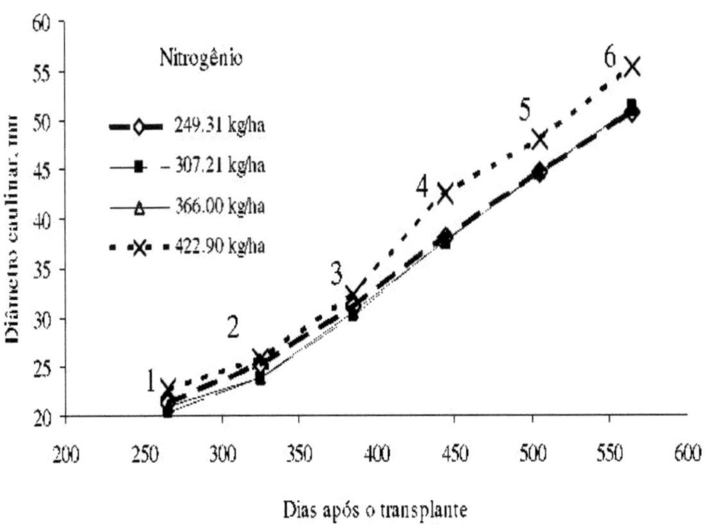

Figure 20 Stem diameter of plants subjected to different water and nitrogen treatments as a function of time. Jequié-BA, 2012.

Table 15. Absolute growth rate (ACR) in (mm.day^{-1}) and relative growth rate (RGR) in (mm.mm^{-1}.day^{-1}) of stem diameter. Jequié-BA, 2012.

Levels -	1---2		2---3		3---4		4---5		5---6	
	TCA	TCR	TCA	TCR	TCA	TCR	TCA	TCR	TCA	TCR
L1	0,06	0,0028	0,10	0,0035	0,12	0,0034	0,11	0,0027	0,10	0,0021
L2	0,06	0,0027	0,10	0,0039	0,13	0,0038	0,11	0,0026	0,11	0,0024
L3	0,05	0,0022	0,11	0,0041	0,12	0,0035	0,12	0,0029	0,11	0,0022
L4	0,05	0,0022	0,11	0,0036	0,17	0,0046	0,09	0,0020	0,12	0,0024
N1	0,06	0,0027	0,10	0,0035	0,14	0,0038	0,11	0,0026	0,11	0,0022
N2	0,06	0,0024	0,11	0,0038	0,13	0,0037	0,12	0,0028	0,11	0,0022
N3	0,05	0,0022	0,11	0,0039	0,15	0,0043	0,09	0,0020	0,11	0,0022

| N4 | 0,05 | 0,0025 | 0,10 | 0,0039 | 0,12 | 0,0035 | 0,11 | 0,0027 | 0,11 | 0,0024 |

4.2.2 Initial production phase

Three stages of cocoa tree development in the field, up to the beginning of the productive phase when the fruits were analyzed in the various aspects proposed in the methodology of this work, are shown in Figure 21.

Figure 21. Evolution of CCN-51 cocoa plants up to the productive stage. A-8 months in the field; B-18 months in the field; C-27 months in the field. Jequié-BA, 2012.

According to the analysis of variance and regression, yield and number of fruits per plant were significantly influenced (0.01 <p <0.05) by the levels of water (L) applied. Nitrogen (N) levels, on the other hand, led to a significant increase (p <0.01) in these characteristics, in addition to the variables number of seeds per fruit, almond weight and number of fruits per plant (Table 16).

The results of the interaction between water laminae and nitrogen doses significantly affected the analyzed characteristics of yield and number of seeds per fruit at the level of (p <0.01) and almond weight and the number of fruits per plant were affected with a probability of 0.01 <p <0.05 (Table 16).

For the variables fruit length, fruit diameter and fruit weight, simulations of regression models were carried out, however, none of these models showed statistical significance of the regression coefficients at the level of (0.01 <p <0.05) and (p <0.01), with these characteristics being better represented by their averages, whose values were 19.41 cm; 85.97mm and 567.34g respectively. Brito and Silva (1983), when studying the fruit of the SIAL-105 clonal cacao tree, found cacao fruit at 180 days of age with a length of 10.37 cm and a diameter of 8.30 cm.

Schroeder (1958), studying the stationary growth of cacao fruit in Costa Rica, observed that dimensional growth in terms of length and diameter follows a sigmoid curve and that growth in length is more prominent than in diameter. However, from 10 cm onwards, diameter growth becomes relatively greater than longitudinal growth, which is associated with major internal tissue changes such as the rapid development of the embryo and the presence of cotyledonary material.

Although it is not necessary to carry out a more detailed study by unfolding the factors water lamina and nitrogen levels because they are quantitative factors (water and nitrogen) at more than two levels, the most appropriate methodology for this study would be through regression analysis using response surfaces, however, in order to verify the influence of one factor on the other within each level, the unfolding was carried out for the characteristics that had a significant effect on the interaction (Table 17 and Table 18).

Table 16. Summary of the analysis of variance and regression of the variables yield (P), fruit length (C), fruit diameter (D), fruit weight (PF), number of seeds per fruit (NSF), almond weight per fruit (PAF) and average number of fruits per plant (NFP), subjected to water levels and nitrogen doses. Jequié-BA, 2012.

FV	GL	Mean Square						
		P, kg.ha^{-1}	C, cm	D, mm	PF, g	NSF	PAF, g	NFP
Blocks	3	121079.31^{ns}	0,46^{ns}	34,74 ^{ns}	3248,43 ^{ns}	41,43 ^{ns}	187,48 ^{ns}	9,91 ^{ns}
Blades (L)	3	216732.75*	2,19^{ns}	40,89 ^{ns}	3049,80 ^{ns}	0,57 ^{ns}	125,87 ^{ns}	23,12*
Nitrogen (N)	3	718877.94**	0,36^{ns}	46,13 ^{ns}	6167,24 ^{ns}	5,30**	467,74**	84,01**
L x N	9	148070.16**	2,91^{ns}	50,02 ^{ns}	8508,48 ^{ns}	4,27**	242,72*	17,36*
Treatments	15	275964.24**	2,26^{ns}	52,82 ^{ns}	6948,50 ^{ns}	83,5**	264,35**	31,84**
Waste	45	51957.79	1,41	61,23	5237,47	22,34	87,27	7,19

58

CV(%)		37,09	6,11	9,10	12,76	11,86	16,67	35,76
L linear	1	50991,87**	0,20 ^{ns}	0,87 ^{ns}	76,87 ^{ns}	2,69 ^{ns}	4,40 ^{ns}	5,25*
L-squared	1	81088,26**	1,42 ^{ns}	0,86 ^{ns}	2109,53 ^{ns}	0,19 ^{ns}	0,43 ^{ns}	7,6**
cubic L	1	38690,09*	0,53 ^{ns}	25,08 ^{ns}	35,72 ^{ns}	7,09 ^{ns}	66,45 ^{ns}	4,45*
N linear	1	371240,38**	0,17 ^{ns}	3,64 ^{ns}	1288,82 ^{ns}	8,46 ^{ns}	$_3$ ^{ns}	55,15*
N squared	1	670,03 ^{ns}	0,09 ^{ns}	6,61 ^{ns}	193,35 ^{ns}	0,24 ^{ns}	1,76 ^{ns}	0,16 ^{ns}
Cubic N	1	175041,41*	0,01 ^{ns}	24,33 ^{ns}	3143,79 ^{ns}	80,06*	336,2**	8,17 ^{ns}

Ns; * ; **; not significant and significant at the 5% (0.01< p< 0.05) and 1% (p < 0.01) probability levels, respectively.

The breakdowns of the variables that showed significance in the interaction are shown in Tables 17 and 18 for the variation of the water laminas within the nitrogen levels used and the nitrogen doses within the water levels.

The variables yield and average number of fruits per plant were significantly affected by the dose of water (N1) (Table 17). This effect is best represented by the linear regression model (Figure 22 A and Figure 22 C). Using these regression models, we found percentage decreases in production and average fruit per plant of 82.69% and 81.52% respectively, considering the highest water table of 2193.91 mm and the lowest of 1384.52 mm. By adding greater amounts of water within these nitrogen levels, the plant's productive capacity is affected. The other characteristics evaluated were not significant for any of the regression models tested.

Only productivity was influenced by the variation in water tables at the nitrogen level (N3) (Figure 22 B) and the quadratic regression model best represents this variation. According to the model, there was a small decrease of 2% when varying the water levels from 1384.52 mm to 2193.91 mm.

Table 18 and Figure 23A, Figure 23B, Figure 23C, Figure 23D and Figure 23E show the variations in nitrogen within each water level. The data best fitted the linear and quadratic models.

The yield of almonds increased linearly with the application of higher doses of nitrogen, with a variation of 82% in lamina L3 and 81.1% in L4 when 45% more nitrogen was added.

The number of seeds per fruit did not increase when the nitrogen doses were increased within the L3 and L4 levels (Figure 21C). The weight of the almonds increased by 34.58 % and 21% when the amount of nitrogen was increased from 318.3 kg ha^{-1} to 580.60 kg ha^{-1} at water levels L3 and L4, respectively (Figure 23).[a]

The number of fruits per plant was the most affected among the parameters, varying from 70.75% at level L3 to 80.55 at L4 when the doses were increased from 318.3 kg ha^{-1} to 580.60 kg ha^{-1} , and at level L4 there was significance for the quadratic model.

Table 17. Summary of the regression analysis of the variables that showed statistical significance as a result of variations in water laminae within nitrogen doses. Jequié-BA, 2012.

Source of variation	GL	Mean Square			
		P, kg.ha^{-1}	NSF	PAF, g	MFP
Δ LN,					
Linear Reg.	1	939434,70**	93,58 [ns]	78,6 [ns]	123,50**

Quadratic Reg.	1	15640.25ns	$140{,}97^{ns}$	$475{,}62^{ns}$	$4{,}62^{ns}$
Δ LN2					
Linear Regulation	1	$155542{,}12^{ns}$	$82{,}25^{ns}$	$237{,}99^{ns}$	$17{,}77^{ns}$
Quadratic Reg.	1	$1656{,}12^{ns}$	$54{,}21^{ns}$	$23{,}32^{ns}$	$0{,}53^{ns}$
Cubic Reg.	1	$23399{,}51^{ns}$	$20{,}83^{ns}$	$164{,}69^{ns}$	$0{,}33^{ns}$
Δ LN3					
Linear Regulation	1	$17205{,}53^{ns}$	$22{,}24^{ns}$	$35{,}98^{ns}$	$6{,}44^{ns}$
Quadratic Reg.	1	$356819{,}71*$	$45{,}71^{ns}$	$150{,}23^{ns}$	$25{,}25^{ns}$
Cubic Reg.	1	$19440{,}85^{ns}$	$14{,}85^{ns}$	$81{,}19^{ns}$	$0{,}33^{ns}$
Δ LN4					
Linear Reg.	1	$24221{,}30^{ns}$	$3{,}70^{ns}$	$16{,}51^{ns}$	$2{,}63^{ns}$

Quadratic Reg.	1	60798,97 ^{ns}	0,52 ^{ns}	4,04 ^{ns}	9,46 ^{ns}
Cubic Reg.	1	4350,56 ^{ns}	0,01 ^{ns}	53,22 ^{ns}	5,78 ^{ns}

Ns; * ; **; non-significance and significance at the 5% (0.01< p< 0.05) and 1% (p < 0.01) probability levels, respectively.

Table 18. Summary of the regression analysis of the variables that showed statistical significance as a result of the variation in nitrogen doses within water levels. Jequié-BA, 2012.

Source of variation	GL	Mean Square			
		P, kg.ha-1	NSF	PAF, g	MFP
Δ_{NL1}					
Linear Reg.	1	548,45 ^{ns}	25,93 ^{ns}	1,57 ^{ns}	0,06 ^{ns}
Quadratic Reg.	1	39451,91 ^{ns}	30,90 ^{ns}	14,18 ^{ns}	3,71 ^{ns}
Cubic Reg.	1	36931,53^{ns}	37,79 ^{ns}	214,83 ^{ns}	0,01 ^{ns}
Δ_{NL2}					
Linear Regulation	1	44458,81 ^{ns}	2,40 ^{ns}	143,2 ^{ns}	21,32 ^{ns}
Quadratic Reg.	1	65655,45 ^{ns}	0,09 ^{ns}	11,46 ^{ns}	17,02 ^{ns}

Cubic Reg.	1	275469,90 ns	96,76 ns	173,75 ns	17,96 ns
Δ NL3					
Linear Reg.	1	1473129,03**	216,28*	1094,14**	147,43**
Quadratic Reg.	1	114015,33 ns	197,08*	154,13 ns	8,12 ns
Cubic Reg.	1	433157,27*	26,63 ns	536,02*	26,22 ns
Δ NL4					
Linear Regulation	1	978497,99**	43,90 ns	237,86 ns	161,60**
Quadratic Reg.	1	186,31 ns	321,82**	617,56*	2,48 ns
Cubic Reg.	1	89008,92 ns	214,43 ns	500,09 ns	4,56 ns

Ns; * ; **; non-significance and significance at the 5% (0.01< p< 0.05) and 1% (p < 0.01) probability levels, respectively.

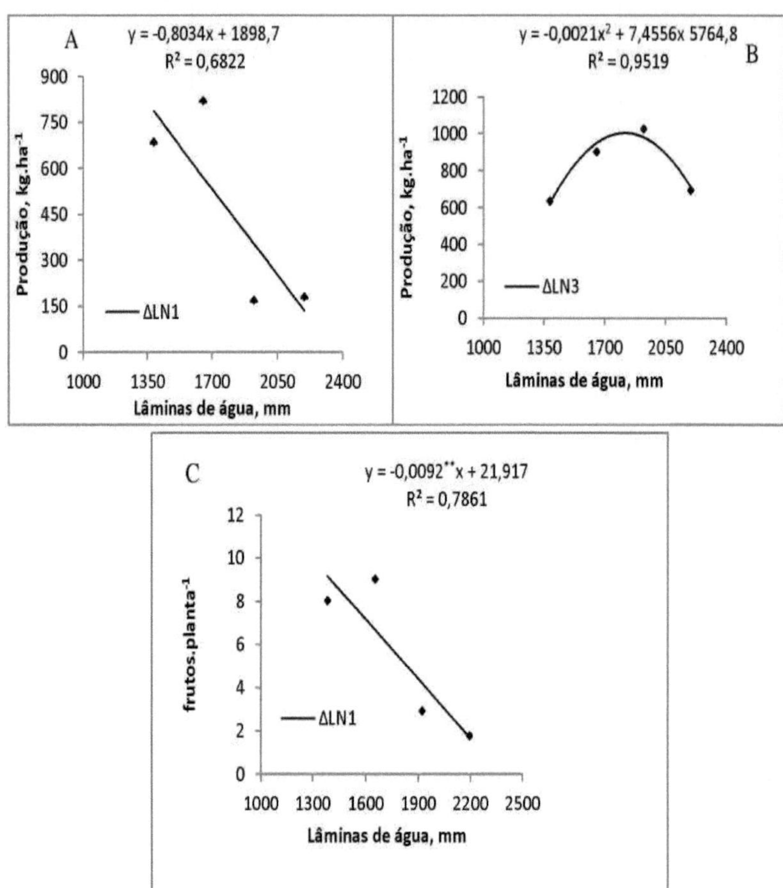

Figure 22 Breakdown for the variable yield and fruit per plant for variations in water laminae and nitrogen doses. Jequié-BA, 2012.

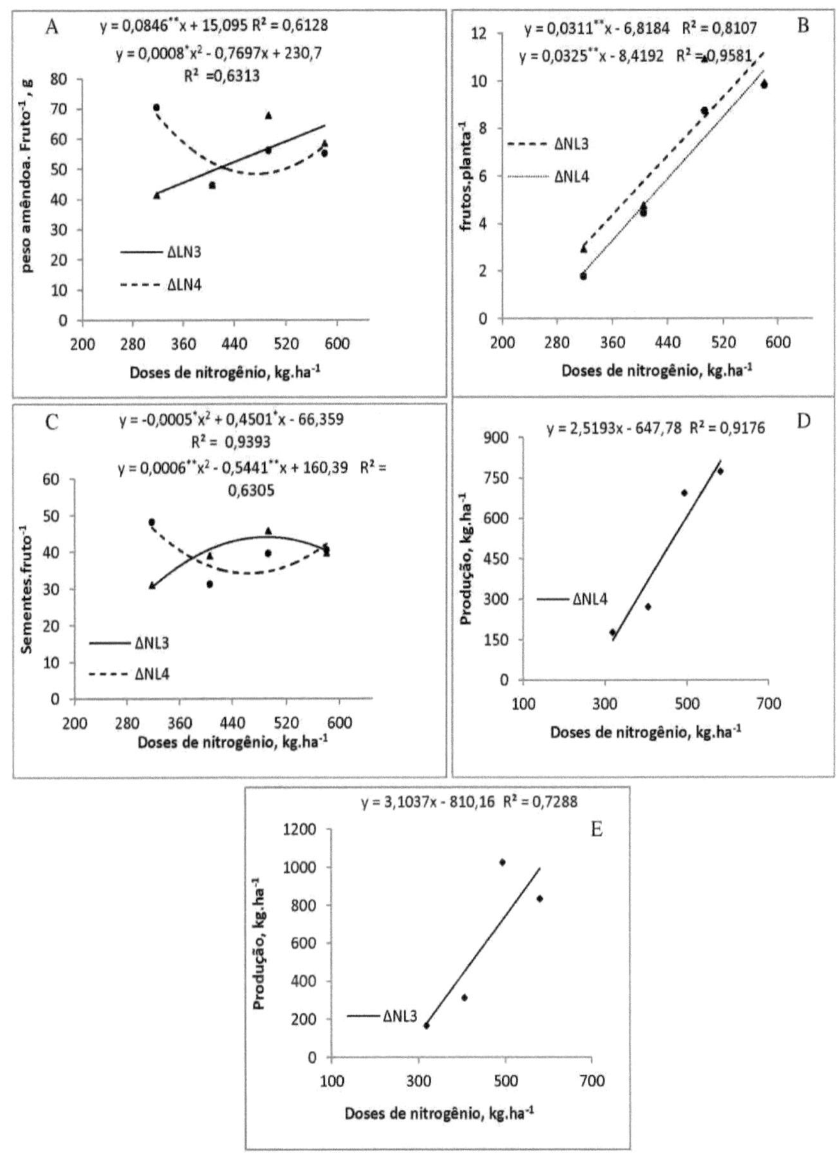

Figure 23 Breakdown for the variable almond weight per fruit, fruit per plant, number of seeds per fruit and yield, for variations in nitrogen doses and water levels. Jequié-BA, 2012.

4.2.2.1. Analysis of the variables number of seeds per fruit, almond weight per fruit and number of fruits per plant.

Tables 19, 20 and 21 show the average values for the variables number of seeds per fruit, almond weight per fruit and number of fruits per plant for the treatments applied in the field. The highest average values achieved for these variables were: 48.25; 70.50 and 11.18 respectively. Coral et al. (1968), analyzing cocoa hybrids, between the third and fourth year of production, IMC 67 x

TSAN 792 found a number of seeds per fruit of 40 and an average weight of seeds per fruit of 42 g, while ICS 1 x IMC showed an average of 42 seeds per fruit and 49 g for the average weight of seeds per fruit. Souza et al. (1996) found an average yield of 16.6 fruits per plant six years after planting.

The percentages of relative growth in the number of seeds per fruit and the weight of the almond in 11.18 cocoa fruits were 55.64% and 69.87% respectively, when using the L4N1 treatment compared to the L3N1 treatment. There was a decrease in the number of fruits per plant (Table 21) when the L4N1 treatment was applied, with a decrease of 528.09% compared to the L2N4 treatment.

Table 19. Number of seeds per fruit for the treatments applied in the field. Jequié- BA, 2012.

Nitrogen -	Lamina				
	1	2	3	4	Average
1	37,19	42,56	31,00	48,25	39,75
2	38,35	37,97	39,00	31,25	36,64
3	43,62	44,22	45,75	39,59	43,29
4	39,23	39,32	39,71	40,53	39,70
Average	39,60	41,02	38,87	39,91	39,85

Table 20. Almond weight (g) per fruit for the treatments applied in the field. Jequié- BA, 2012.

Nitrogen	Lamina				Average
	1	2	3	4	
1	56,50	63,69	41,50	70,50	58,05
2	52,10	56,81	44,75	44,63	49,57
3	62,22	62,97	67,68	56,18	62,26

| 4 | 54,06 | 52,71 | 58,51 | 55,15 | 55,11 |
| Average | 56,22 | 59,05 | 53,11 | 56,61 | 56,25 |

Table 21. Number of fruits per plant for the treatments applied in the field. Jequié-BA, 2012.

Nitrogen	Blade				
	1	2	3	4	Average
1	8,03	9,03	2,93	1,78	5,44
2	7,15	6,10	4,78	4,45	5,62
3	7,15	9,98	10,93	8,73	9,19
4	8,20	11,18	9,93	9,83	9,78
Average	7,63	9,07	7,14	6,19	7,51

4.2.2.1.1. Regression models

Tables 22, 23 and 24 show the significance values of the model coefficients in relation to Student's t-test, all the variables included in the model showed
results were highly significant at an error of 1%, with the exception of the coefficients for N, N^2 and N^3, which were significant at an error of 5%. All the coefficients could therefore be included in the regression models. The coefficients of determination (R^2) were relatively low at 0.58, 0.54 and 0.60 for the variables number of seeds per fruit, kernel weight per fruit and number of fruits per plant respectively, but the significance of the coefficients validates the regression models.

Table 22. Significance of the regression model coefficients for the response surface number of seeds.fruit^{-1}. Jequié-BA, 2012.

Elements of equation	Estimated parameters	Pr > |t|
Intercept	451,72	" " " =£=£ 0,0017

N	-2,93	0,0035**
N^2	0,0067	0,0032**
N^3	-0,000005	0,0031**

**; significance at 1% (p < 0.01) probability level.

Table 23. Significance of the coefficients of the regression model for the response surface weight of the kernels.fruit^{-1} . Jequié-BA, 2012.

Elements of the equation	Estimated parameters	Pr > |t|
Intercept	919.79255	0.0006
N	-6.07707	0.0012**
N^2	0.01384	0.0012**
N^3	-0.00001024	0.0013**

**; significance at 1% (p < 0.01) probability level.

Table 24. Significance of the regression model coefficients for the response surface number of fruits.plant^{-1} . Jequié-BA, 2012.

Elements of equation	Estimated parameters	Pr > |t|
Intercept	173.46757	0.0114*

N	-1.01054	0.0323[*]
N^2	0.00216	0.0435[*]
N^3	-0.00000159	0.0446[*]
L	-0.02147	0.0002[**]
L x N	0.00004260	0.0006[**]

*; **; significance at the 5% (0.01< p< 0.05) and 1% (p < 0.01) probability levels, respectively.

4.2.2.1.2. Response surface

The characteristics that presented regression equations with significant coefficients have their response surfaces shown in Figures 24, 25, 26 and 27.

The response surface (Figure 24) showed a cubic model as significant for the variable number of seeds per fruit, with nitrogen being the only factor to influence this variable. The response surface generated shows a linear increase in the number of seeds as a function of the doses of nitrogen applied. According to the regression model used, the number of seeds per fruit increased by 16.78% as the amount of nitrogen increased from 318.30 kg ha^{-1} to 580 kg ha $^{-1}$

$$\hat{P} = 451{,}71 - 2{,}93N + 0{,}0067N^2 - 0{,}000005N^3$$

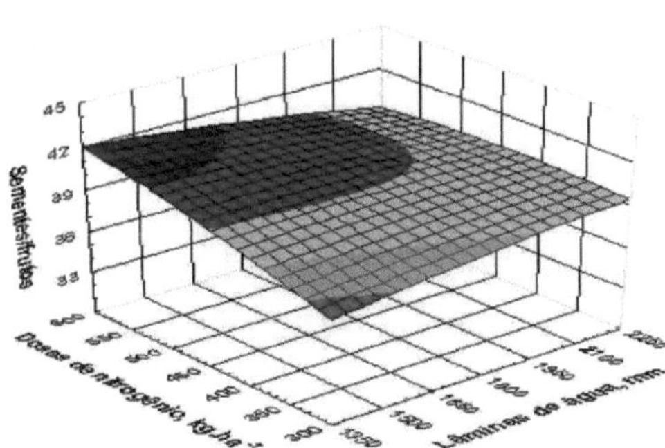

Figure 24 Average number of seeds in CCN-51 clonal cacao fruits as a function of water laminae and nitrogen doses. Jequié-BA, 2012.

Nitrogen caused a linear increase in the weight of the almonds from the lowest dose of 318.30 kg ha^{-1} to the highest of 580 kg ha^{-1} . The regression model used showed a 47% gain in weight for a 45.12% increase in nitrogen. According to the response surface (Figure 25), as nitrogen increases, the gain in weight of the cocoa beans is very small; this is due to the small variation in the weight of one unit of beans from the CCN-51 clone, which is approximately 2g on average.

$$\hat{Y} = 919,79 - 6,08N + 0,014N^2 - 0,00001N^3$$

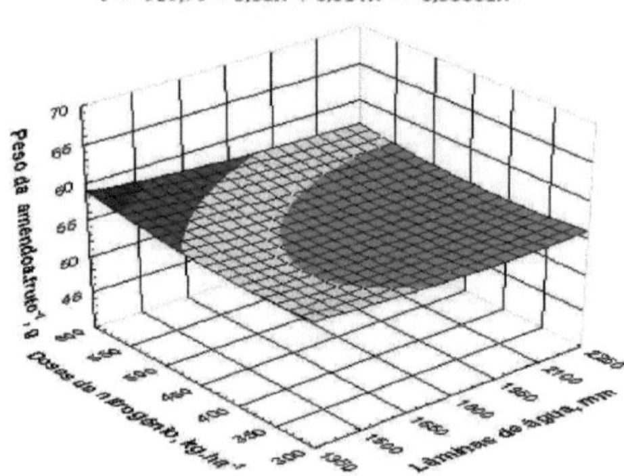

Figure 25 Average weight (g) of almonds from CCN-51 clonal cacao fruit, as a function of water laminae and nitrogen doses Jequié-BA, 2012.

The effect of the interaction between the factors water table and nitrogen was significant for the number of fruits per plant with probability (P<0.01) and the regression model used (Figure 26) is considered highly significant, according to the significance of the coefficients (Table 24).

The response surface shows a linear trend in relation to the doses of nitrogen and a cubic trend in relation to the water levels applied (Figure 26).

The highest average number of fruits per plant was found for treatment L3N3, whose average value was 11.01 fruits per plant. Lower values were found by Souza Junior (1997), Gramacho et al. (1992) and Silva Neto (2001) for traditional regions; these researchers found an average production of 4.5 fruits per plant from the fourth year of planting[-1], but this production increases with planting age, reaching a balance from the tenth year onwards with a production of 28 to 35 fruits per plant[-1].

Leite (2006) found an average production per plant considering only those that came into production of 3 fruits per plant[-1], with the CCN 51 and PH 16 clones standing out with 4.9 and 3.5 fruits per plant[-1], respectively, at 21 months from planting. The production data obtained showed that the cacao tree was up to 2 years earlier than in traditional regions. In these regions, according to Seagri (2012), the cocoa tree starts producing from the age of 3.

$$\hat{Y} = 173,46 - 1,01N + 0,022N^2 - 0,0000016N^3 - 0,02L + 0,000043LN$$

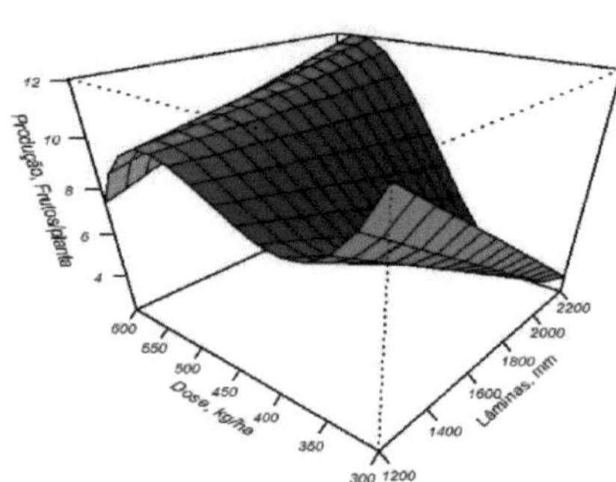

Figure 26 Fruits.plants[-1], of clonal cacao CCN-51, as a function of water laminae and nitrogen doses. Jequié-BA, 2012.

4.2.2.2. Analysis of eight months of production

Table 25 shows the equivalence between the productivity of CCN-51 clonal cacao in kg ha^{-1} and its equivalent in arroba.ha^{-1} and in fruit ha^{-1} . The highest yield of 1025.69 kg ha^{-1} obtained with the L3N3 treatment had its equivalent of 68.38 arroba.ha^{-1} which corresponds to using 15.20 fruits to obtain one kilogram of almonds for sale, a value which is compatible with that estimated to produce one kilogram in traditional regions.

Figure 27 shows productivity in the agricultural units of measurement that are most commonly used in the marketing of cocoa at national and international level. The graph shows the lowest yields for treatments L3N1 and L4N1, with yields of 11.13 arroba. ha^{-1} and 11.87 arroba. ha^{-1} , respectively.

In the irrigated perimeters of the São Francisco Valley, during a 10-year production cycle, an average annual productivity of 200 arroba. ha^{-1} (approximately 3000 Kg) is considered. This productivity is based on the average production of CEPLAC's experimental areas (Begiato et al., 2009). This productivity has been achieved over the years, as according to Mi (2006) the cocoa tree reaches its productive optimum at around 35 years of age. According to this author, economic production begins in the fifth year after planting. Corroborating this idea, Dias et al. (1998) reported that the average yields showed an upward trend with the advancing age of the cocoa trees and great variability between years.

Table 25. Cocoa tree yield in kg.ha^{-1} and its equivalent in arroba.ha^{-1} and fruits.ha^{-1} . Jequié-BA, 2012.

TREATMENTS	Water	Nitrogen	Kg.ha^{-1}	arroba.ha^{-1}	fruit.ha^{-1}
L1N1	1384,52	318,30	683,04	45,54	11471,66
L2N 1	1653,22	318,30	819,28	54,62	12900,26
L3N1	1922,52	318,30	166,90	11,13	4185,80
L4N1	2193,91	318,30	178,07	11,87	2542,91
L1N2	1384,52	405,80	503,02	33,53	10214,49
L2N2	1653,22	405,80	503,59	33,57	8714,46
L3N2	1922,52	405,80	312,79	20,85	6828,71
L4N2	2193,91	405,80	272,67	18,18	6357,27
L1N3	1384,52	493,10	637,18	42,48	10214,49
L2N3	1653,22	493,10	902,82	60,19	14257,43
L3N3	1922,52	493,10	1025,69	68,38	15614,60
L4N3	2193,91	493,10	693,99	46,27	12471,68
L1N4	1384,52	580,60	655,78	43,72	11709,60
L2N4	1653,22	580,60	843,37	56,22	15971,75
L3N4	1922,52	580,60	833,92	55,59	14185,99
L4N4	2193,91	580,60	774,93	51,66	14043,14

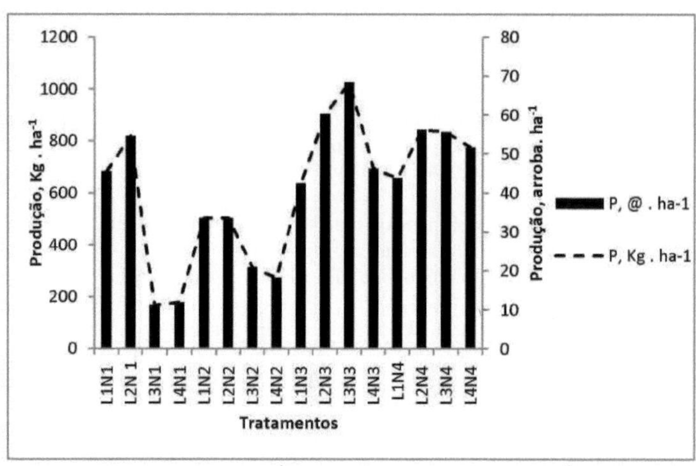

Figure 27 Cocoa tree production in kg.ha^{-1} and its equivalent in arroba.ha^{-1} . Jequié-BA, 2012.

27.2.2.2.1.Production function

Three regression models were tested for the variable almond production in kg ha^{-1} , according to the methodology used in the paper. Tables 26, 27 and 28 show the significance of the coefficients of the models that were analyzed. Model 1 showed the effect of L, L^2 and L^3 to be non-significant at the 5% probability level using Student's t-test (Table 26), with a coefficient of determination (R^2) of 0.86.

Model 2, described in the methodology, whose coefficients are shown in Table 27, proved to be the most reliable regression model as it showed all the coefficients to be significant at the 5% probability level using Student's t-test, with the exception of the coefficient of L which remained in the model as L^2 was significant, showing a coefficient of determination (R^2) of 0.83, i.e. a good fit with the data analyzed.

Regression model 3 was tested, with a second degree format constructed by removing the N effect[3] even though this effect was significant. The results of the significance of the coefficients are shown in Table 28. However, this new regression model showed all the coefficients to be non-significant at the 5% probability level using the t-test, with the exception of the interaction whose p-value was equal to 0.0019 and therefore significant at the 5% probability level. This model showed a coefficient of determination (R^2) of 0.66.

Table 26. Significance of the coefficients of regression model 1 for the yield response surface. Jequié-BA, 2012.

Elements of equation	Estimated parameters	Pr > |t|
Intercept	1323,84	0,9247^{ns}
N	-143,75331	0,0007*

	0,31591	* 0,0011*
N^2		
N^3	-0,00023366	0,0011*
L	34,11272	0,1251^{ns}
L^2	-0,01955	0,1210^{ns}
L^3	0,00000348	0,1377^{ns}
L x N	0,00370	0,0007*

Ns; * ; **; non-significance and significance at the 5% ($0.01 < p < 0.05$) and 1% ($p < 0.01$) probability levels, respectively.

Table 27. Significance of the coefficients of regression model 2 for the production response surface. Jequié-BA, 2012.

Elements of equation	Estimated parameters	Pr > |t|
Intercept	20331	0,0013*
N	-143,75331	0,0008*
N^2	0,31591	* 0,0012*
N^3	-0,00023366	0,0012*

L	1,20192	0,4516[ns]
L^2	-0,00086238	0,0458[*]
L x N	0,00370	0,0008[*]

Ns; * ; **; non-significance and significance at the 5% (0.01< p< 0.05) and 1% (p < 0.01) probability levels, respectively

Table 28. Significance of the coefficients of regression model 3 for the production response surface. Jequié-BA, 2012.

Elements of equation	Parameters estimated	Pr > |t|
Intercept	762,96690	0,6909[ns]
N	-5,81146	0,1954[ns]
N^2	0,00084621	0,8472[ns]
L	1,20192	0,4885[ns]
L^2	-0,00086238	0,0654[ns]
L x N	0,00370	_ _ _ _ # 0,0019[*]

Ns; * ; **; not significant and significant at the 5% (0.01< p< 0.05) and 1% (p < 0.01) probability levels, respectively.

Using the selected regression model shown in Figure 28, we calculated the values of water depth and nitrogen dose that provide the maximum physical yield, which are 1926.23 mm and 560.70 kg ha^{-1} , respectively. For this optimum combination, a value of 1649.23 kg ha^{-1} or 109.95 arroba.ha^{-1} of dry cocoa beans was estimated. This was 46.75% higher than the maximum observed in the field of 1025.69 kg ha^{-1} or 68.38 arroba.ha^{-1} for treatment L3N3.

The optimum economic combination of water and nitrogen found from the regression equation (Figure 28) was 1832.69 mm and 552.86 kg ha^{-1} , respectively, providing 1637.40 kg ha^{-1} or 108.60 arroba.ha^{-1} of dry cocoa beans.

27.2.2.2.2. Response surface

The response surface methodology was used to identify the water level and nitrogen dose that provided the maximum physical yield, using the partial derivatives in relation to the factors present in the production function.

The curvature profile on the response surface (Figure 28) shows that there is a point of maximum production of cocoa for sale which increases with the increase in the level of nitrogen applied and increases up to an amount of water that the plant can absorb and from which all the excess water causes a drop in production levels.

The response surface shows a tendency towards a linear increase in relation to the doses of nitrogen, i.e. the more nitrogen added to the soil, in the same proportion there is a greater yield and a cubic decrease in yield in relation to the water laminas applied.

$$P = 20331 - 143.75N + 0.32N^2 - 0.00023N^3 + 1.2L - 0.00086L^2 + 0.0037LN.$$

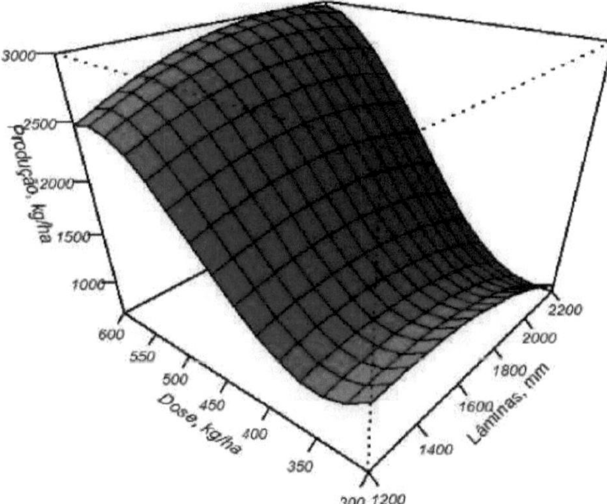

Figure 28 Yield of clonal cacao CCN-51, as a function of water laminae and nitrogen doses.

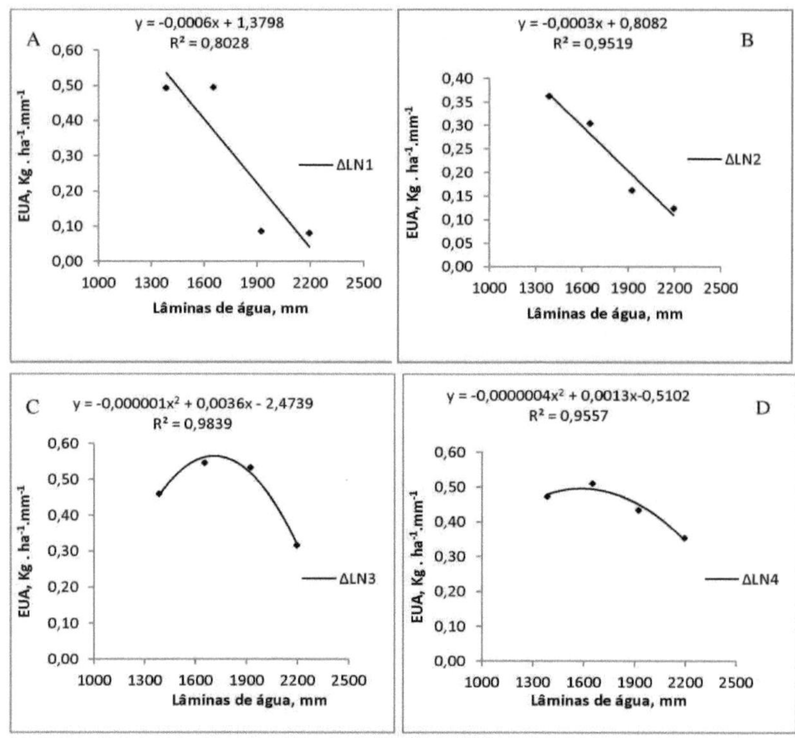

Figure 29 Water use efficiency (WUE) as a function of water laminae at nitrogen levels, A-N1,B-N2,C-N3 and D- N4. Jequié-BA, 2012.

4.2.3. Water use efficiency (USA)

Water use efficiency (WUE) decreased with increasing water rates as crop yield did not increase with increasing water rates (Figure 29).

The lowest water table used in the experiment (1384.52 mm) proved to be more efficient both in the physical production of cocoa beans and economically, because for a 36.89% reduction in water compared to the highest table (2193.91 mm), at levels N1 and N2 there was an increase of 88.0% and 62% in EUA (Figure 29A and Figure 29B).

The highest US value was 0.55 kg ha^{-1} .mm^{-1} for the treatment with a water table of 1922.52 mm and a nitrogen dose of 493.10 kg ha^{-1} (L3N3), producing a total of 1056.75 kg ha^{-1} of dried cocoa beans (Figure 30). This result indicates that, under the best conditions, 1 kg of cocoa beans would require 18,136 liters of water, respectively. The maximum efficiency that occurred with variations in water at nitrogen levels N3 and N4, using the regression equations, was 0.77 kg ha^{-1} mm^{-1} and 0.56 kg ha^{-1} mm^{-1} for 1800 mm and 1625 mm (Figure 29C and Figure 29D), respectively.

Coelho et al. (2005) agree that the efficiency of water use can be increased by reducing the amount of water applied so as not to drastically reduce productivity. Most of the time, an increase in US can be achieved by decreasing the amount of water applied (Letey, 1993).

Other researchers have found similar results where yields were increased by reducing the amount of water applied.

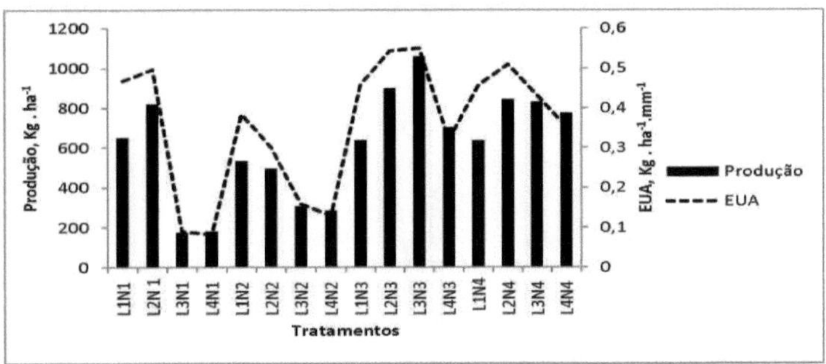

Figure 30 Water use efficiency (WUE) and production in the treatments adopted in the experiment. Jequié-BA, 2012.

4.2.4. Nitrogen use efficiency (NUE)

There was a tendency for EUN to decrease as the amount of nitrogen increased, with the dose of 318.30 kg ha⁻¹ showing a percentage 45.18% lower than the dose of 580.6 kg ha⁻¹ , where EUN was 46.57% higher. According to the regression models (Figure 29A and Figure 29B), there was always a tendency for EUN to decrease with the application of higher doses of nitrogen at lower water levels. Figures 31 C and 31 D explain the evolution of EUN through quadratic regression models, where these models show that the highest value of EUA occurred for the doses of 516.67 kg ha⁻¹ and 741.67 kg ha⁻¹ both showing the same efficiency with a value of 1.55 kg of grain (kg of N) .⁻¹

The highest EUN found with field data was 2.58 kg of grains (kg of N)⁻¹ for a dose of 318.30 kg ha⁻¹ of nitrogen and a water table of 1653.22 mm, (L2N1), these values show that for each kg of N applied there was an increase of 2.58 kg of grains in the production of cocoa beans. For the lowest EUN of 0.55 kg of grains (kg of N)⁻¹ with a dose of 318.30 and a water table of 1922.52 mm, there was a gain of 0.55 kg in production, i.e. 78.68% less in the production of cocoa beans (Figure 32).

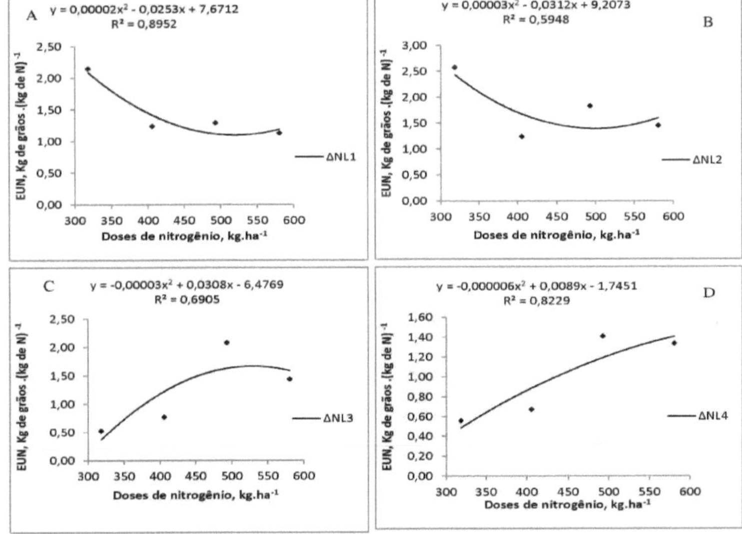

Figure 31 Nitrogen use efficiency (NUE) as a function of nitrogen doses at water levels,A- L1,B-L2,C-L3 and D- L4. Jequié-BA, 2012.

EUN tended to decrease as the water table increased, showing that these variables have a direct relationship with nitrogen use, affecting the efficiency of using this input; this is probably due to the leaching of these nutrients through the application of larger amounts of water.

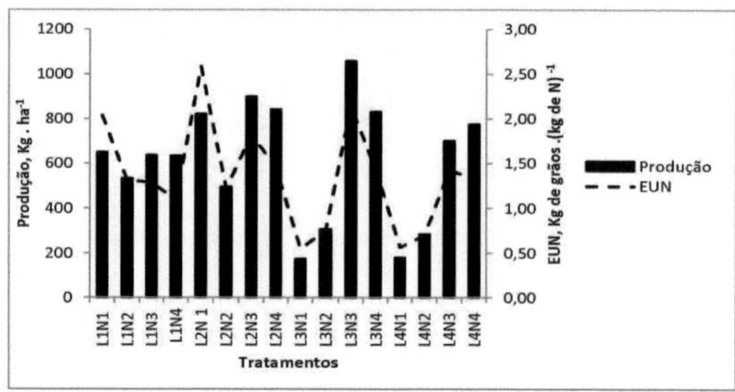

Figure 32 Nitrogen use efficiency (NUE) and production in the treatments adopted in the experiment. Jequié-BA, 2012.

4.6. Simplified economic analysis

Table 29 shows a simplified economic analysis for the combination of water rates and nitrogen doses, as well as the estimates of these factors that promote maximum physical and economic yield.

Table 29 shows that the maximum physical and economic yield estimated from the production function occurred with the combination of LF = 1,567.30 mm of water and NF = 395.79 kg ha⁻¹ of N, and LE = 1,505.50 mm of water and NF = 394.00 kg ha⁻¹ of N, with yields of R$ -2368.80 and R$ -2330.65 respectively.

These quantities showed a negative net income for all the possible combinations shown in Table 29; this is mainly due to the fact that this analysis was carried out when the cocoa was producing for approximately ten months. During this period, the costs of setting up the irrigation system and growing the crop in the field were higher than the gains from production.

From the results in Table 23, it can be seen that the combination that resulted in the lowest loss of revenue for this production period was treatment L3N3, with a 63% higher yield than the combination that resulted in the highest revenue deficit, which was L3N1.

Table 29. Yield and net revenue for the various combinations of water laminae and nitrogen doses. Jequié-BA, 2012.

Treatments	Water	Nitrogen	Yield, kg.ha⁻¹	Revenue,R$
L1N1	1384,52	318,30	683,04	-1970,93
L2N1	1653,22	318,30	819,28	-1452,16
L3N1	1922,52	318,30	166,90	-4932,08
L4N1	2193,91	318,30	178,07	-5049,14
L1N2	1384,52	405,80	503,02	-3097,13
L2N2	1653,22	405,80	503,59	-3266,21

L3N2	1922,52	405,80	312,79	-4405,92
L4N2	2193,91	405,80	272,67	-4783,02
L1N3	1384,52	493,10	637,18	-2629,95
L2N3	1653,22	493,10	902,82	-1455,13
L3N3	1922,52	493,10	1025,69	-1004,53
L4N3	2193,91	493,10	693,99	-2859,94
L1N4	1384,52	580,60	655,78	-2749,15
L2N4	1653,22	580,60	843,37	-1970,04
L3N4	1922,52	580,60	833,92	-2190,30
L4N4	2193,91	580,60	774,93	-2663,07
PHYSICS	**1926,23**	**560,7**	**1649,23**	**1989,50**
ECONOMIC	**1832,69**	**552,86**	**1637,4**	**2008,518**

CONCLUSIONS

The seedlings of the CCN-51 clone showed good adaptation to the existing conditions, with the number of leaves standing out.

The water levels and nitrogen doses used in this study influenced plant growth in terms of height and stem diameter, while the interaction between the water and nitrogen factors had no influence on these characteristics.

Water levels, nitrogen doses and the interaction between them had a positive influence on the production of dry almonds and the number of fruits per plant of clonal cacao CCN-51.

The treatment used in the field with 1922.52 mm of water and 493.10 kg ha^{-1} of nitrogen was the one that produced the highest yield of cocoa beans for sale with 1025.69 kg ha^{-1} .

The physical optimum combination of water depth and nitrogen dose was 1926.23 mm and 560.70 kg ha^{-1} of nitrogen, which gave a yield of 1649.23 kg ha^{-1} of dried cocoa beans.

The treatment with 1922.52 mm of water and 493.10 kg ha^{-1} of nitrogen obtained the highest efficiency of water use corresponding to 0.55 kg ha^{-1} mm^{-1} and the treatment with 1653.22 mm of water and 318 kg ha^{-1} of nitrogen obtained the highest value of efficiency of nitrogen use with 2.58 kg of grains (kg of N)$^{-1}$.

In the initial production phase, the optimum economic combination of water table and nitrogen dose was 1,832.69 mm and 552.86 kg ha^{-1} which gave a net income of R\$ 2008.518 for a production of 1,637.4 kg ha^{-1} .

RECOMMENDATIONS FOR OPTIMIZING THE GROWTH AND PRODUCTION OF IRRIGATED COCOA TREES

In general, in order to obtain better results in terms of the growth and production of cocoa trees subjected to irrigation, we can make some recommendations:

It is essential to implement initial shading with corn to help the seedlings adapt to the field, but avoid intercropping the cacao tree with permanent crops to avoid competition for nutrients in the soil. In semi-arid or arid climates, avoid drastic pruning during the production period to prevent the fruit from burning. Whenever possible, place windbreaks around the plantations to prevent the cocoa leaves from drying out due to the action of the wind and the plant from falling over.

Avoid excess water during irrigation management and be careful of leaks in drip hoses and pipes during the cocoa tree's growth phase. Managing the application of water based on the crop's evapotranspiration is essential for the plant to grow normally and produce satisfactorily.

In the first year in the field, intensify the nitrogen fertilization by applying 6 to 8 g of N.plant-1 weekly to accelerate growth in height and stem diameter via fertigation or apply 15 g of N.plant manually^{-1} every 15 days. From the third year in the field, when the cocoa begins to produce, apply 10 to 12 g of N.plant^{-1} weekly to speed up production or 20 g of N.plant^{-1} every 15 days.

From the fifth and sixth year onwards, apply 16 to 18 g of N.planta^{-1} weekly to increase production or 30 g of N.planta^{-1} every 15 days. It is important that the cacao tree is fertilized with micronutrients, especially sources containing iron, magnesium and zinc. The amount will depend on a more accurate analysis of the soil to check for deficiencies.

With regard to potassium, in the first year in the field you can apply 5 g of N.plant^{-1} , in the productive phase you should increase this dose to 10 g of N.plant^{-1} and from the 6th year onwards it is recommended that you increase the dose to 15 g of N.plant^{-1} . Phosphorus should be applied to the hole one month before the seedlings are planted, following the guidelines mentioned above.

Phosphorus should be applied one month before planting in holes measuring 0.40 x 0.40 x 0.40 m, or 50 x 50 x 50 m; 70 g (hole)$^{-1}$ of FTE BR12, a chemical compound made up of micronutrients, added to 140 g (hole)$^{-1}$ of MAP (52% P2O5 + 11% N), i.e. 72.8 g (hole)$^{-1}$ of phosphorus pentoxide (P2O5) plus 15.4 g (hole)$^{-1}$ of nitrogen (N2), adding 3 liters of goat manure to this mixture as an organic fertilizer.

BIBLIOGRAPHICAL REFERENCES

ABDOELLAH, S.; NOTORADININGRAT, T. Effect of Al/(K+Ca+Mg) ratio on growth of cocoa seedling. **Pelita Perkebunan**, v.9, n.1, p.23-28. 1993.

ABDUL-KARIMU, A.; ADU-AMPOMAH, Y.; FRIMPONG, E. B. Field evaluation of agronomic characters of some selected cocoa hybrids in a marginal area of Ghana. In: INTERNATIONAL COCOA RESEARCH CONFERENCE, 14, 2003. Accra, Ghana, **Abstract...**2003, Section 3, CD.

ADU-AMPOMAH, Y.; FRIMPONG, E. B.; ADOMAKO, B.; ABDUL-KARIMU, A. Investigation into the Use Of the Crinkle Leaf Mutant as a Low Vigour Rootstock for High Density Planting in Cocoa. **International Workshop on Cocoa Breeding for Improved Production Systems.** Accra, Ghana, 2003, p.145-149.

AGUIAR, J. V. **The production function in irrigated agriculture.** Fortaleza: University Press, 2005. 196p.

ALKANANI, T.; MAQCKENZIE, A.F.; BARTHAKUR, N.N. Soil water and ammonia volatilization relationships with surface applied nitrogen fertilizer solutions. **Soil Science Society American Journal**, v.55, p. 1761-1766, 1991.

ALMEIDA, R. L. S.; CHAVES, L. H. G. Growth of cocoa seedlings irrigated by micro-sprinklers. **Engenharia Ambiental: Pesquisa e Tecnologia,** v. 7, n. 2, p. 284-293, 2010.

ALMEIDA, A. A. F.; VALLE, R. R. Ecophysiology of the cacao tree. **Brazilian Journal of Plant Physiology,** v.19, n.4, p.425-448, 2007.

ALVES, R. C.; DEL PONTE, E. M. **Vassoura-de-bruxa.** Available at: http://www6.ufrgs.br/agronomia/fitossan/fitopatologia/ficha.php?id=273: Accessed on February 23, 2010.

ALVIM, P. T. **Cacao: yesterday and today.** Ilhéus: CEPLAC, 1972. 83p.

ARCE, M.P. MINISTERIO DE AGRICULTURA PROGRAMA PARA EL DESARROLLO DE LA AMAZONIA PROAMAZONIA **"MANUAL DEL CULTIVO DEL CACAO".** Peru, 2004.83p.

AUGUSTO, S. G. **Complementary irrigation in the different phenological stages of the cacao tree (*Theobroma cacao* L.).** Vinosa: UFV, 1997, 120p. Thesis (Doctorate in Irrigation and Drainage).

BEGIATO, G. F; SPERS, E. E; CASTRO, L. T; NEVES, M. F. Analysis of the agro-industrial system and attractiveness of the San Francisco Valleys for irrigated cocoa farming. **Custos e @gronegócio,** v. 5, n. 3, p. 55-87. 2009.

BENINCASA, M. M. P. **Analysis of plant growth**. Jaboticabal: Funep, 2003. 41p.

BERNARDO, S.; SOARES, A.A.; MANTOVANI, E.C. **Manual de Irrigado.** 8.ed, Vinosa: UFV, 2006. 625p.

BOARETTO, A.E.; MURAOKA, T.; TREVELIN, P. **Efficient use of nitrogen in conventional fertilizers.**Informagoes agronómicas, v.120, p.13-14.2007.

BORGES, A. L.; SILVA, D. J. **Fertirrigacao em frutiras tropicais,** Cruz das Almas: Embrapa Mandioca e Fruticultura Tropical. v. 1, ed.1, p. 1-137, 2002.

BRASILEIRO DE FRUTICULTURA, 20, 2008, Vitória-ES, **Anais** Vitória-ES, 2008. CD ROM.

BRAUDEAU, J. **El cacao.** Barcelona, Translated by Cardona, A.M. H. 1.ed. Barcelona: Editorial Blume. 1970, 297 p.

BRITO, I.C.; SILVA, C.P. Biometric measurements of cacao fruit during development. **Sitientibus,** v.3, n.2, p.59-66,1983.

BURRIDGE, J. C.; LOCKARD R. G.; ACQUAYE D. K. The Levels of Nitrogen, Phosphorus, Potassium, Calcium and Magnesium in the Leaves of **Cacao** (Theobroma **Cacao** L.) as affected by Shade, Fertilizer, Irrigation, and Season. **Botany Company**, 28ª ed, p.401418, 1964.

CANNELL, M.G.R. Dry matter partitioning in tree crops. IN: CANNELL, M.G.R., JACKSON, J.E. (Eds.), **Attributes of Trees as Crop Plants**. ITE/Wilson, UK, 1985. p. 160193.

CASTRO, P. R. C.; KLUGE, R. A. **Ecophysiology of tropical fruit trees**. Sao Paulo. 1998. 111p.

CEPEC/CEPLAC.　　　**General characteristics of cocoa**. Available at: http://www.ceplac.gov.br/radar/cacau.htm Accessed on: September 10, 2009.

CHEPOTE, R.E.; SODRE, E.; REIS, E.L.; PACHECO, R.G.; MARROCOS, P.C.L.; SERODIO, M.H.C.L.; VALLE, R.R. **Recomendacöes de corretivos e fertilizantes na cultura do cacaueiro no sul da Bahia**. ed.2 Ilheus: CEPLAC/CEPEC. 2005.36p.

CHRISTIANSEN, J. E. **Irrigation by sprinkling**. Berkeley: University of California, 1942. 124 p.

CODEVASF. **Cocoa production chain: investment opportunities in cocoa farming in the Vale do Sao Francisco and Parnaíba**. Brasilia, 2009. 33 p.

EXECUTIVE COMMISSION OF THE COCOA PLANTATION PLAN (CEPLAC). **Cocoa farming. Market information**. Available at: http://www.ceplac.gov.br/paginas/infomercado/. Accessed on: October 10, 2011.

CRUZ, P. At a time of progress and diversification, Bahian fruit-growing is already harvesting up to macä. http://www.seagri.ba.gov.br/noticias.asp?qact=view&exibir=clipping¬id=9538, 05 Jun. 2010.

COELHO, E. F.; COELHO FILHO, M.A.; OLIVEIRA, S.L. Irrigated agriculture: irrigation and water use efficiency. **Bahia Agrícola**, v. 7, n. 1, p. 57-60, 2005.

CORAL, F.J.; CIONE, J.; IGUE, T. Preliminary studies on the behavior of hybrid cacao tree progenies in the ecological conditions of the Ribeira Valley. **Bragantia**, v.27, p. 63-65,1968.

DIAS, L.A.S.; RESENDE, M. D. V. Experimental breeding. In: DIAS, L.A.S. **Melhoramento genético do cacaueiro**. Viçosa: FUNAPE-UFG, 2001. p.439-492.

DIAS, L. A. S.; SOUZA, C.A.S.; AUGUSTO, S.G.; SIGUEIRA, P.; MULLER, M.W. Performance and temporal stability analyses of cacao cultivars in Linhares, Brazil. **Plantations, Recherche, Développement,** v. 5, n. 5, p. 343-355, 1998.

DOOREMBOS, J.; KASSAM, A. H. **Effect of water on crop yields.** Campina Grande: UFPB, 1994. 306p. (FAO Studies: Irrigation and Drainage, 33).

DOOREMBOS, J.; PRUITT, W.O. **Guidelines for predicting crop water requirements**. Rome: FAO, 1997. 179 p.

DUKE, J. A. The quest of tolerant germplasm. In: YOUNG, G. **Crop tolerance to subtropical land conditions.** Madison. American Society Agronomial Special Symposium, 1978, v.32, p.1-16.

DUKE, J.A. *Theobroma caçao* **L**. Handbook of Energy Crops.1983

EFRON,Y.; TADE, E.; EPAINA, P. A cocoa growth mutant with a dwarfing effect as rootstock. In: **International Workshop on Cocoa Breeding for Improved Production Systems,** 14, 2003, Accra, Ghana, 2003, p. 132-144.

EMBRAPA. **Citrus Production System for the Northeast.** Available at: http://sistemasdeproducao.cnptia.embrapa.br . Accessed on June 10, 2009.

FAO. Foundation Agricultural Organization. Rome: FAOSTAT Database Gateway. Available at:< http://www.apps.fao.org>Accessed **September 2011.**

FRAZÁO. D. A. C.; COSTA, J. D.; CORAL, F. J.; AZEVEDO, J. A.; FIGUEREDO, F. J. C. Influence of seed weight on the development and vigor of cocoa seedlings. **Revista Brasileira de Sementes**, v.6, n. 3, p. 31-40, 1984.

FRIZZONE, J. A. **Crop response functions to irrigation.** Piracicaba: ESALQ/USP, 1993. 42p. (Didactic Series, 6).

GAGNÉ, S. Cacao: The Essence of Chocolate. Available at:

http://www.stevegagne.com/more.php. Accessed on: **July** 28, **2008.**

GRAMACHO, I. C. P.; MAGNO, A. E. S.; MANDARINO, E. P.; MATOS, A. **Cultivation and processing of cocoa in Bahia.** Ilhéus, Ceplac. 1992. 124 p.

KLAR, A.E. A **água no sistema solo-planta-atmosfera.** 2ª ed. Sao Paulo: Nobel, 1988.

HARDY, F. **Manual de cacao.** Inter-American Institute of Agricultural Sciences. Turrialba, Costa Rica. 1961. 439p.

HEXEM, R.W.; HEADY, E.O. **Water production function for irrigated agriculture.** Ames: The Iowa State University Press,1978. 215p.

LEITE, J. B. V. **Cocoa trees: Propagation by stem cuttings and planting in the semi-arid region of the state of Bahia.** 2006. 75 f. Thesis (Doctorate) - UNESP, Jaboticabal, 2006.

LETEY, J. Relationship between salinity and efficient water use. **Irrigation Science,** v.14, p.75-84, 1993.

LOBÃO D. E.; SETENTA, W. C. Cacao - cabruca: history and characterization of a sustainable agroforestry system of proven efficiency. In: BRAZILIAN CONGRESS ON AGROFORESTATION SYSTEMS, 4, 2002, Ilhéus. Proceedings..., **Ilhéus,** 2002.

LOPES, A.S. **Manual de fertilidade do solo.** Translated and adapted from the original: Soil fertility manual. Potash Phosphate Institute, 1978. Sao Paulo: ANDA/POTAFOS, 1989. 153p.

LUZ, E.D.M.N.; BEZERRA, J.L.; OLIVEIRA, M.L.; RESENDE, M.L.V. Cacao tree diseases. In ZAMBOLIM, L.; VALE, F.X.R. **Controle de doencas de plantas:** Grandes culturas.Vigosa:UFV,1997.cap.13,p.611-655.

MARROCOS, P. C. L.; SODRÉ, G. A. Cacao tree seedling production system. In: ENCONTRO NACIONAL DE SUBSTRATOS PARA PLANTAS, 4, 2004, Vigosa. **Abstracts** Nutrition and fertilization of plants grown in substrate. Vigosa: UFV, 2004. p. 283-311.

MARSCHNER, H. **Mineral nutrition of higher plants.** 2. ed. London: Academic Press,1995. 889 p.

MINISTRY OF NATIONAL INTEGRATION. **Cocoa agronomic studies - 2nd progress report.** Preparation of the socio-technical-economic and environmental feasibility study, and EIA/RIMA for the Cruz das Almas irrigated perimeter of the Sertao de Pernambuco project, located in the municipality of Casa Nova-BA. Bahia, 2006.

MOLL, R. H.; KAMPRATH, E. J.; JACKSON, W. A. Analysis and interpretation of factors which contribute to efficiency of nitrogen utilization. **Agronomy Journal,** v. 74, n. 3, p. 562564, 1982.

MOOLEEDHAR V.; LAUCKNER, B. Effects of spacing on yield in improved clones of *Theobroma cacao* L. **Tropical Agriculture,** ed.67, Trinite-Et-Tobago, v.67, n.4, p. 376-378, 1990.

MORAIS, F.I., SILVA, L.F.; MARINHO, A.H.; PINHO, A.F.S. Effects of foliar fertilization and substrates on the growth of cocoa plants. **Revista Theobroma,** v.9, p.163-171, 1979.

MULLER, M. W.; BIEHL, B. Changes in the photosynthetic capacity of cacao leaves (*Theobroma cacao* L.) influenced by light intensity during the life span. In: INTERNATIONAL COCOA RESEARCH CONFERENCE, 11., 1993, Yamoussoukro, Côte d'Ivoire, **Abstract...** .1993, p. 634-643.

NAKAYAMA, L.H.I.; SOARES, M.K.M.; APPEZZATO-DA-GLÓRIA, B . Contribution to the anatomical study of the leaf and stem of the cacao tree (Theobroma Cacao L.). **Scientia Agrícola.** v.53, p. 73-73, 1996.

NAMALIU, Y.; DANIEL, R. **Integrated pest and disease management for sustainable cacoa production.** Australia, 2008. 37p.

NANETTI, D. C.; SOUZA, R. J.; FAQUIN, V. Effect of nitrogen and potassium application, via fertigation, on the pepper crop. **Revista Brasileira de Olericultura**, v.13, p. 843845, 2000.

NETAFIM. **Cocoa**. Available at: **Error! The hyperlink reference is not valid.** Accessed on July 21, 2008.

OETTERER, M. Technologies for obtaining cocoa, cocoa products and chocolate. In: OETTERER, M.; REGITANO D'ARCE, M. A.; SPOTO, M. H. F. **Fundamentos de Ciências e Tecnologia de Alimentos**. 1.ed. Barueri, SP: Manole, 2006, v. 1, p. 1-50.

OLIVEIRA, L.S. **Response functions of sweet corn to the use of irrigation and nitrogen**. Viçosa, MG: UFV, 1993. 91p. (Doctoral thesis).

PEIXOTO, C.P.; PEIXOTO, M.F.S.P. **Dynamics of plant growth (Basic Principles)**. Cruz das Almas. 2004. 20p.

PROAMAZONIA. MINISTRY OF AGRICULTURE - Amazon Development Program. **Cacao cultivation manual**. Ministry of Agriculture, Peru. 2004

PURDY, L.H.; SCHMIDT, R.A. Status of cacao witches'broom: Biology, Epidemiology, and Management. **Phytopathol**, ed.34, p.573-594, 1996.

REED, C.F. **Information summaries on 1000 economic plants**. Typescripts submitted to the United States Department of Agriculture (USDA). Washington, 1976.

SANTOS, L. S.; RIBEIRO, V. G. Evaluations of cloned cacao trees CCN-10, CCN-51, PS-1319 AND PH-16: from seedling production to grafting, in the semi-arid region of Bahia. In: CONGRESSO BRASILEIRO DE FRUTICULTURA, 20, 2008, Vitória-ES, **Anais** Vitória-ES, 2008. CD ROM.

SARTORI, P.H. **Efficiency of nitrogen and sulfur use by sugarcane (first and second regrowth) in a conservation system (without burning)**. Piracicaba 2010. 112p. (Doctoral thesis.)

SCHROEDER, C.A. **Observation on the growth of the cacao fruit**. In: 7ª Conferencia Interamericana de Cacao. Colombia, Bogota, p. 381-394, 1960.

SEAGRI (Secretariat of Agriculture, Irrigation and Agrarian Reform). **Cocoa Culture**. Available at: http://www.seagri.ba.gov.br/cacau1.htm Accessed on: March 1, 2012.

SILVA NETO, P. J. (Coord.) **Cocoa production system for the Brazilian Amazon.** Belém, PA. CEPLAC, 2001. 125 p.

SIRQUEIRA, P.R.; MULDER, W.T.J.; SOUZA, C.A.S. **Fertirrigação do cacaueiro no estado do Espirito Santo**. Available at: http://www.ceplac.gov.br . Accessed on: March 23, 2011.

SIRQUEIRA, P. **The importance of irrigation for cocoa in Linhares**. Available at: http://www.ceplac.br . Accessed on July 21, 2008.

SMITH, G. E. (Ed.). **Cacao cultivation in the state of Espírito Santo**. Ilhéus: CEPLAC/CEPEC, 1990. 52 p.

SOUZA, A.G.C.; SILVA, S.E.L.; SOUSA, N.R. **Avaliagao do desempenho do cacau-do- peru.** EMBRAPA/CPAA, Manaus, p.1-2, 1996.

SOUZA, C. A. S.; AGUILAR, M. A. G.; SONEGHETI, S.; BOONER, E. P.; CAO, J. R.; PINTO, D. G. Production of clonal cacao seedlings in polyethylene bags with different substrates and forms of fertilizer. In: CONGRESSO BRASILEIRO DE FRUTICULTURA, 20, 2008, Vitória. **Anais...**, Vitória, 2008. CD- ROM.

SOUZA JÚNIOR, J. O.; CARMELLO, Q.A.C. Forms of fertilization and doses of urea for clonal cocoa seedlings grown in substrate. **Revista Brasileira de Ciencia do Solo,** v. 32, n. 6, p. 2367-2374, 2008.

SOUZA JUNIOR, J. O. **Edaphoclimatic factors influencing the productivity of cocoa trees grown in the south of Bahia, Brazil**. Master's thesis, UFV. Vigosa, MG, 1997. 146 p.

TEAL, F.; VIGNERI, M. **Production changes in Ghana cocoa farming households under market reforms**. Ghana, 2004.28p.

UNITED STATES DEPARTMENT OF AGRICULTURE. **Cacao tissue culture**

protocol book. Pennsylvania: State University, USA. 2003. 32p.

VERNON, A.J.; SUNDARAM, S. MURRAY, D.B.; JONES, E.; QUESNEL, V.C.; CHALMERS, W.S.; FORDHAM, R.; ITON, E.F. Current cocoa research in Fiji. **Proceedings**, International Cocoa Research Conference, 4, St. Augustine, 1972, p. 689-693.

WILHELM, W. W.; MCMASTER, G. S. Importance of the phyllochron in studying development and growth in grasses. **Crop Science**, v.35, n.1, p.1-3, 1995.

ZOBEL, R.W.; KINRAIDE, T.B.; BALIGAR, V. Fine root diameters can change in response to changes in nutrient concentrations. **Plant and Soil**, v. 297, p. 243-254, 2007.

ZUIDEMA, P.A.; LEFFLAAR, P.A.; GERRITSMA, W.; MOUMER, L.; ANTEN, N.P.R. A physiological production model for cocoa (Theobroma cacao); model presentation, validation and application. **Agricultural Systems,** v.84, p.195-225, 2005.

Printed by Books on Demand GmbH, Norderstedt / Germany